THE WOMEN ARE MORE BEAUTIFUL THAN ANY I HAVE EVER SEEN

I0895498

Sidestone Press

The women are more beautiful than any I have ever seen

Heinrich Schliemann's travels through Spain in 1859

edited by:
Wout Arentzen & Maaike van Asch

The Schliemann Diaries 4

Published by Sidestone Press, Leiden
www.sidestone.com

ISBN 978-90-8890-693-0 (softcover)
ISBN 978-90-8890-694-7 (hardcover)
ISBN 978-90-8890-695-4 (PDF e-book)

Lay-out & cover design: Sidestone Press
Photograph cover: oilpainting: *Un paisaje. Recuerdos de andalucía*. 1860. By Carlos de Haes (1826-1898), Museo del Prado, Madrid. Copy from: www.museodeplrado.es.

Background cover: A1 diary; Gennadius Library, Athens

The Schliemann Diaries 4 | www.sidestone.com/schliemann/

Contents

Preface

> *Published diaries are often hypocritical beasts, craftily appealing to the voyeur in each of us whose dearest wish is to get to know other people's deepest, uncensored feelings. "This is for my eyes only" .... Don't try to fool us!: if the diarist really hadn't wanted others to discover his most intimate secrets, he would never have kept his diaries in the first place.*
>
> De Rek, Wilma. Schrijf het toch op (Publish and be damned). *De Volkskrant* 2 January 2016

This is the fourth part in the series *The Schliemann Diaries*. The set-up of this series is simple: the publication of all Schliemann's diaries in the form of a transcript, an English translation and an introduction to the text. The aim of this series is to unlock to the public the multi-lingual diaries in English. The original text can be read and interpreted with the help of the transcript. The English translation and the introduction should make the diaries accessible to everybody. The idea is to have the numbering of the diaries (A1, A2, ...) coincide with the numbering of the series. This is diary A4 and so it is part 4 of this series. We hope that every book will form the basis for a further in-depth study of the life and world of Heinrich Schliemann.

Schliemann was one of the first archaeologists who opted for a multidisciplinary approach to his work. We follow the same approach in the transcription and translation of his diary.

The diary was in the first instance transcribed and translated by Maaike van Asch. We would like to thank Iñaki do Campo Gan for assistance in the translation work and for bringing us into contact with Walid Aissaoui who translated the Arabic passages in Schliemann's diary. For the transcription and translation of the Arabic passages we thank Abdurraouf Oueslati.

We further acknowledge the assistance and permission for publication from the Gennadius Library in Athens, and in particular Natalia Vogeikoff-Brogan. Thanks are also due to Dr. Armin Jähne, Dr. Wilfried Bölke, Gerhard Pohlan and Dr. Reinhard Witte from the archives of the Heinrich Schliemann Museum in Ankershagen, Germany.

Gail Neffenger and Dr. Edmund F. Bloedow translated the transcript into English, Robert Bathgate translated the explanatory notes and chapter 3, while Myra Wilkinson edited the English translation of the diary. Christo Thanos prepared the manuscript for publication.

Our special thanks are due to Dr. Edmund F. Bloedow for his introductory essay and to Dr. Armin Jähne for the translation and annotation of Ekatherina Schliemann's letters.

Wout Arentzen
Maaike van Asch
November 2016

Figure 1: Heinrich Schliemann 1861 – Gennadius Library

Edmund F. Bloedow (1931-2019)

A Background to Schliemann's 1858-1859 Diary

Schliemann's idiosyncratic personality turned his archaeological career, as brilliant as it was, into an unnecessarily controversial legend, which has produced a separate body of research that stretches from before his first excavations in 1870, when he was 48 years of age, down to this day. His personal life prior to his career as an archaeologist is in many respects also the stuff of legend, especially his acquisition of enormous wealth within the space of about 11 years (1847-1858), chiefly in Russia. This too has produced its own body of research, which likewise continues down to the present.

Towards the end of the latter phase Schliemann embarked on a grand tour to the Orient (1864-1866). At the end of this tour occurred what one can only construe as an incidental side-trip of two days. Within the overall context of Schliemann's life this side-trip cannot merit more than an insignificant footnote. Indeed, for the longest time it did not even receive the slightest notice in the large body of already existing research.

This little side-trip was to Canada, more especially to Montreal (2 November, 1865). In a brief discussion of this episode, I concentrated on the possible reasons for Schliemann's private visit to the family of James Ferrier.[1] There is, however, considerably more to the story, which opens a significant further window on an essential part of Schliemann's life.

Here it is relevant to pause to consider the sources of our information for Schliemann's personal life. These consist chiefly of thousands of letters making up his correspondence, as well as the various diaries of his numerous travels.[2] This raises a significant question. Was it ever Schliemann's intention to publish the information in this body of material, or that it should be published after his death? There are some very personal details that one can scarcely imagine Schliemann would ever have wished to be published. This, accordingly, raises a question about all the rest. At this point in time, however, the question may appear to be essentially irrelevant. Since the material is now publicly available, it seems entirely legitimate to use it in research. And certainly much of it has done a great deal to illuminate Schliemann's personality, and his personality in many respects significantly illuminates his activity as an archaeologist.

1 E. F. Bloedow, "Heinrich Scliemann and James Ferrier. A Noteworthy Side-Trip to Canada," in Pierre Senay (ed.), Mêlanges *Pierre Rodrigue Brind'Amour I.* Cahiers des Études Anciennes 37 (Trois Rivières, PQ, 2001), 9-26.

2 These are found primarily in the Gennadius Library in Athens today.

In respect of our current discussion, the principal factor is Schliemann's relationship with women.[3] We shall note a number of features that reappear again and again – not least the fact that many of the women involved went by the name of 'Sophie'.

Sophie Schwarz

In 1823, when Schliemann was only one year old, his father, Ernst Schliemann (1780-1870), became the pastor in the village church in Ankershagen. The hamlet had a small population of slightly more than 1000, and lay in a rather remote and backward region of northern Germany. By this time there were five children.[4] It became a very rocky marriage, thanks in particular to the pastor's far from ethical behaviour. Within a short time Ernst Schliemann had hired a maid, Sophie Schwartz, ostensibly to help his wife, but his intentions proved to be otherwise. He soon had an affair with her. In such a small community the news quickly spread. By 1830 Luise Schliemann was able to get rid of Sophie Schwartz. But it took an immense toll on her, for she died in 1831, shortly after giving birth to her ninth child. She was only 38.

The matter was investigated by the church authorities. The pastor was found guilty, but succeeded in dragging the episode out until 1838, when he was forced to abdicate his post and leave town.

But meanwhile, shortly after the death of Schliemann's mother, his father re-instated Sophie Schwarz, and they lived together in a common-law relationship. The parishioners and the villagers were outraged. The following Sunday 200 of them staged a demonstration with clanging of pots and pans.

In 1832 Ernst Schliemann arranged to have Heinrich leave idyllic and legend-rich Ankershagen and go to live with his brother Christian Ludwig Friedrich Schliemann, pastor of the church in Kalkhorst.

This whole affair had a deeply disturbing effect on Heinrich at a very impressionable age, to which he gave expression a number of times in print. He wrote, for example, "my father, who did not fail to observe my deep depression, sent me off for two years to his brother [Friedrich]... the pastor of the church in Kalkhorst". This took place in the period 1832-34.

Minna Meincke

A major factor in Schliemann's life was the local Ankershagen legend of the heroic figure Henning Brandenkirl. He recounts this in great detail on a number of occasions in later life especially in his autobiography of 1881. It becomes clear that this legend greatly fired his imagination.[5]

---

3 Here I shall draw chiefly on my study, "Die anderen 'Sophies' in Schliemanns Leben," *Das Altertum* 47 (2002), 283-305 – namely, because this study is not so readily available and also because it was written in German.

4 In 1813 Ernst Schliemann had married Luise Sophie, Theresa Bürger (1793-1831) (13 years younger than himself), the daughter of the Rektor and later mayor of Sternberg.

5 See further below.

Another episode is his famous boyhood dream he claimed to have had of one day excavating Troy. A number of modern sceptics have questioned this, maintaining that it is a late invention which Schliemann wove into his autobiography. While this is an admittedly attractive thesis, it is based essentially on arguments *ex silentio*. What is not adequately explained is that at the age of 59 Schliemann recounts this dream in all its elaborate detail, at the beginning of his most famous book, *Ilios*, published in 1881. While this anecdote does not particularly enhance the book, its inclusion provides significant insight into his personality.[6]

Closely associated with Schliemann's boyhood dream of one day excavating Troy was his unforgettable recollection of his first love – with Minna Meincke, a farmer's daughter from the neighbouring village of Zahren. In his autobiography at the beginning of his *Ilios* he writes:

> *My mother's death coincided with another terrible misfortune. As a result, all our acquaintances turned their backs on us, and cut off all contact. The others did not cause me much grief, but the fact that I should no longer see the Meincke family, that I should have to break off all contact with Minna, that I should never see her again – this was a thousand times more painful than my mother's death. I soon recovered from the latter, but not from the disaster of losing Minna. We had sworn our love to each other, and that one day we would excavate Troy together. But now my whole future looked dark and bleak. All the mysterious marvels of Ankershagen – indeed of Troy itself lost their attraction. In later life I suffered many afflictions in different parts of the world, but not one of these caused me a fraction of the pain that at the tender age of nine years separation from my little bride engulfed me. Bathed in tears I stood for hours each day alone in front of the painting of Olgartha von Schröder and in deep sorrow thought of the happy days I had spent in Minna's company.*[7]

Moreover:

> *Some days before my departure from Neu-Strehlitz, on Good Friday 1836, I inadvertently met Minna Meincke once more, in the house of the court musician C. E. Laue.*[8] *I had not seen her for more than five years. Never shall I forget this meeting, the very last that was to be granted us! She was now fourteen years old, and had really grown since I had last seen her. She was wearing a black dress, and the very simplicity of her attire seemed to enhance her captivating beauty. When our eyes met, we both broke into a flood of tears, and unable to utter a word we fell into each other's arms. Again and again we tried to speak, but our emotions were so intense we simply could not. But soon Minna's parents entered the room, and we were forced to break up. It took a long time before I regained my composure. Now I was sure that Minna still loved me, and this thought fired my ambition. From that moment I felt an unbounded energy and the sure confidence within myself that through untiring determination I would succeed in the world and*

---

6 H. Schliemann, Ilios. The City and Country of the Trojans (London 1881).

7 *Illios*, 6-7.

8 On this family and its connection with Russia, see E. Meyer, *Heinrich Schliemann, Kaufmnn und Forscher* (Berlin 1969), 58-59.

demonstrate that I was worthy of Minna. The only thing I pleaded before God was that He would prevent Minna from becoming married until I had acquired an established position.[9]

During his stint in Amsterdam Schliemann did not forget Minna. By plunging himself into most exacting work he wanted to recover from his misery:

I entertained the wish to show myself worthy of Minna, which created and nurtured within me an unconquerable determination. And so I applied myself with the greatest diligence to study English.[10]

Then, in St. Petersburg, shortly after he had settled there and saw the prospect of further commercial success, his thoughts returned to Minna and marriage.

I had scarcely succeeded in reaching a completely independent position but what I promptly wrote to the friend of the Meincke family, C. E. Laue, in Neu-Strehlitz. I described my present circumstances, and asked him immediately to seek Minna's hand on my behalf. How great was my shock when after a month I received his devastating answer that just several days hitherto Minna had married another man.[11]

At the time, this disappointment struck me as the cruelest fate that could ever overtake me. I felt completely incapable of doing anything, indeed it left me sick. Endlessly I recalled everything that had taken place between Minna and me in our first childhood – all our sweet dreams and massive plans, as I visualised how brilliantly they could now have been realised. But, must I now contemplate carrying these out by myself? I then also hurled the most vile accusations against myself – that before leaving for St. Petersburg I had not sought her hand. But again and again I had to tell myself that, had I done so, I would have made myself look ridiculous; In Amsterdam I was nothing; that I was wholly dependent on the mood of my employer; that I had no guarantee that I would fare any better in St. Petersburg, indeed I could end up a complete failure. I thought it just as impossible that Minna could be happy were she to marry another man as I could were I to marry another woman. Why should cruel fate snatch her away from me now, when I, after vying for her for sixteen years, believed that I had succeeded in my suit? It had indeed happened to both of us as so often occurs in dreams: we imagine ourselves compulsively pursuing someone, but then, as soon as we think we are about to catch him, he slips out of our grasp. At the time I thought that I would never be able to recover from the pain of losing Minna; but time which heals all wounds, finally brought its healing influence to bear on my spirit, and even though I pined over her loss for years, I was gradually able to pursue my business activities without distraction.[12]

---

9 *Ilios*, 7-8.

10 *Ilios*, 11-12.

11 She married a farmer, Georg August Friedrich Richter (he lived nearby) on 26 November, 1847. She was 26, he was 45. After a marriage of 35 years, he died on 13 July, 1882. She outlived him by 28 years, and died on 15 may, 1910, at the age of 88 (Meyer, 1969, 135).

12 *llios*, 14-15.

It is extraordinary to find such an autobiography at the outset of a book like *Ilios*. Indeed, a number of eminent scholars and leading publishers sought to dissuade Schliemann from including it.[13] The story in itself is of course not unbelievable, but what is arresting is that at the age of 59 and in the face of counsel against the move, Schliemann should have insisted on publishing it as he did. But this in fact fits in with other aspects of his relationship with women – as the following emphatically illustrate.

Sophie Schliemann

Long before Schliemann received the news of Minna's marriage, i.e., just a few years after they parted at their last meeting on that Good Friday, we find him attracted to another Sophie – this time his cousin, the daughter of his uncle in Kalkhorst.[14] In his letter to his sisters of 20 February, 1842, he recounts how he had met her again in 1841:

> *How nature has transformed the girl in eight years! She is so slim and elegant. The natural and artless gracefulness of her movements, the delicacy and beauty of her facial features, her lovely sparkling eyes that she fixed upon me – enchanted me, and I am ready to maintain that the engaging imagination of a poet at the pinnacle of his creative powers never saw such an ideal of noble beauty as stood before me in timeless grace. She too seemed profoundly moved, for I saw how her lovely cheeks soon reddened, then soon turned pale … I had to tell her my entire life history, with all its adventures. It was a moving 'Good bye' as I pressed glowing kisses on her lips.*[15]

This was not the last encounter between Schliemann and Sophie, as we shall see.

Sophie Hecker

Let us recall first his musings 33 years later over the devastating discovery that Minna had married but a month before his marriage proposal sent via Laue: "But time, which heals all wounds, eventually worked its wonder in restoring my

---

13 Thus Rudolf Virchow, John Murray, the publisher of Ilios, and Madame E. Egger, who translated the *Ilios* into French. Meyer (1969, 136-137), on the other hand, supported Schliemann in his intention to publish it.

14 One might well ask just how genuine Schliemann's feelings towards Minna were in 1847 as described in his Ilios. The fact is that at the end of 1846 he was wealthy enough to interrupt his business activity and embark on a tour of Europe, and wealthy enough to make a marriage proposal in respect of Minna. It might therefore be regarded as somewhat surprising that in his correspondence between 1836 and 1846 we do not find any reference to Minna, not to mention a letter either from or to her. In his Diary of this tour he does express his feelings as he approaches the coast of his fatherland, but there is no reference to Minna. Nor does he express any intention to disembark and to see her, although he was so close. Instead, he proceeds on his journey to see other places. Was he oscillating once more in his feelings?

15 GL Sch BB 1, 18. (GL = Gennadius Library, Sch BB = Heinrich Schliemann Papers, series BB: original letters.)

spirit". We in fact discover that 'time' was a very quick 'healer'. We see Schliemann attracted to another young lady – within the space of a year. This time it is a young lady from a German family living in St. Petersburg – Sophie Hecker.

Here we encounter a phenomenon that was to characterise all his subsequent relations with women – hesitation and insecurity, and otherwise an inability to make up his mind. In simple and almost naive manner he, on the one hand, asks his sisters for advice, whereas, on the other, he seems fully ready to plunge into a decisive decision. To his sisters he writes that he is:

> *the husband of the most adorable creature one could ever imagine. You write, 'Decide only after serious and mature reflection!' But, my God, I do not pretend to judge on the basis of the virtues of an angel, and I settled on my future wife on our first meeting. She struck me forthwith as noble, and I saw only her virtues … I am on the cusp of sheer delight. What a recompense after so much suffering! She is masterful at playing the piano, and speaks three European languages fluently … Sophie is very frugal, and so we could become wealthy.*[16]

This was the first time Schliemann became engaged to be married. But he soon discovered that he had been too hasty in his decision – at least in respect of certain aspects of his personality which had hitherto remained buried within him but now rose to the surface. The old vice of distrust and jealousy drove him to break off his engagement but a year after his passionate description in his letter to his sisters. It seems that Sophie showed too great an interest in a young officer. Schliemann's own description of her behaviour runs:

> *Alas, your father's endearing blessing was not destined to be fulfilled … On the 15th of November we were together at an event when Sophie showed particular interest in an officer. This made me angry and jealous, and so I left the party. And this is probably for the best, for Sophie is still very young and immature.*[17]

At the same time he demonstrates a certain rashness, as he continues:

> *Marrying is not difficult here,*[18] *especially for someone in my position: I have an unsullied reputation, no debts, a cash flow of 10,000 Taler, and the prospect of earning an additional 16,000 this year. If I wait until fall, I could marry the richest and most beautiful young lady.*[19] *Regrettably I have no patience, and I intend, soon after Easter, to marry the young lady who caught my fancy before I met Sophie … She is a beautiful, very clever Russian lady, with little or no wealth.*[20]

---

16 Emil Ludwig, *Schliemann of Troy. The Story of a Goldseekeer* (London 1931), 82.

17 Ludwig 1931, 83.

18 But it was admittedly very difficult to find the right individual, not to mention the right decision in connection with the right person.

19 One notes the attributes which he mentions (this is not the only such instance). Naturally, such a woman need not in herself be in fact the 'right' person. It is therefore worth mentioning that when he did marry, his decision was based on neither wealth nor beauty.

20 Ludwig 1931, 83.

Ekaterina Lyshin

His intention to marry after Easter (1847) did not come about. It seems that this 'clever and beautiful Russian lady' was less enthusiastic about him than he was about her. And so for the time being he concentrated his energies once more on his business activities. These, in 1850 (1851) and 1852, took him for the first time to the United States, where he discovered an immeasurably vast and interesting new world.[21]

About five months before leaving the United States he decided to approach Sophie Hecker once more. From California he sent her a written marriage proposal. But at the same time, he did not put all his *eggs* in one basket. At the same time he sent a similar proposal to Ekaterina Lyshin.[22]

True, Sophie and Ekaterina were good friends, but neither was particularly enthralled by Schliemann's tactics. One is tempted to speculate whether there may have been some kind of strategy behind his approach, and what this may have been. Since both women were annoyed, one may ask whether either was seriously interested in his proposal. In any event, in the 'competition' that seems to have followed, Ekaterina emerged as 'victor'.[23] And in any event, he appealed once more to his sister:

> *Ah yes, sister, there are enough brides, in fact hundreds. You can help me in choosing. I myself am blind. Passion robs me of my vision. I see only the virtues, never the weaknesses of women.*[24]

Heinrich Schliemann and Ekaterina Lyshin were married on 12 October, 1852 (i.e., two months after his return from the United States). The wedding took place in St. Isaac's Cathedral in St. Petersburg. Schliemann was 30 years old, Ekaterina 20. The marriage was not a happy one. From the very beginning Heinrich and Ekaterina quarrelled. It seems that the epitome was reached in 1863, when Schliemann decided to close his business in St. Petersburg for good and retire, and settle in Paris. He sought to persuade Ekaterina to join him, but she categorically refused. He made heroic efforts to persuade her – without success. It would appear that Schliemann was more interested in having his children around him than he was in Ekaterina. Finally, he decided to get a divorce. The prospects for such in Russia at the time were non-existent, not to mention Ekaterina's stubborn opposition. The most likely prospect was in the United States. And so it was to the USA that he betook himself, in 1867, to test the waters. But not until 1869 did he, after strenuous efforts, achieve success.

---

21 He left St. Petersburg on 10 December, 1850, and returned on 4 August, 1852.

22 In a letter to his sisters four years earlier, when he already had an eye for Ekaterina, he appealed to them to come to stay with him for a time in St. Petersburg or in Moscow, especially to help him in choosing a wife. Among other things, he writes: "My friend Givago is worth millions. This is the same individual who was with me years ago and was the reason for my coming to Russia. His wife is without comparison and the most delightful individual I have ever seen in my life, and his niece Ekaterina is an angel in virtue and beauty … Ekaterina is 16 years old" (Ludwig 1931, 86).

23 If Sophie in fact thought further about Schliemann's proposal and decided to withdraw from the drama?

24 Ludwig 1931, 86.

Meanwhile, on 6 January 1868, on his 46th birthday, Schliemann found himself in Washington, DC, and thought again of Sophie Schliemann (i.e., Cousin Sophie) whom he had kissed so passionately a quarter of a century earlier. She had meanwhile, now in her fifties, written to him via Paris:

> *Many thanks, indeed, for your love, my dear Henri! I was not able to write immediately, but my thoughts and prayers followed you. What do you think of the idea of taking your money there and buying property here? As soon as you have purchased property here, you can be sure that land will not lose its value. In the summer I should very much like to take a trip with you by land. Now my dear, beloved Henri, keep well.*
>
> *Love from your dear Sophie Schliemann.*[25]

This letter awaited him two months later on his return. His answer gives the impression of a lack of virtually any feeling. It seems that he was not even aware of what he expressed in (indeed, much less between) the lines:

> *Dear Sophie,*
>
> *You express the wish to go on a trip with me! But my dear, I confess that your morals are far too virtuous for me. At the time of my departure from Boltenhagen you refused to embrace me! Since you constantly refused to take my arm, how could you go on a trip with a man of the world? Most gladly would I travel with a woman of the world, but I cannot imagine anything more boring that travelling with a saint, who is much more suited for a Nunnery than for the great theatre of the world.*[26]

Although Schliemann posted this letter, Sophie was no longer able to read it, for she died on the same day as Schliemann penned the lines. When he received the news of her death from his aunt, he was totally devastated, and expressed his feelings in a letter to his aunt:

> *I can never forgive you for not telling me about her illness! How gladly would I have dropped everything here and rushed to her side! How gladly would I have fetched the best doctors in Hamburg or Berlin! How gladly would I have kept vigil at her bedside – and perhaps could have rescued her! It was no sensual love, there was nothing calculated in my attraction to her or bound me to her angelic personality – rather, the purest Platonic attachement, the noblest empathy.*

By referring to her as a 'saint,' he was of course only trying to tease her.

> *How gladly would I have gone with her on a trip around the world! Ah, what would I have given to have a photograph of Sophie's coffin! … Despite all my inexpressible sorrow at Sophie's death, I must none the less be glad that I no longer belong to a country where there does not appear to be any human sympathy! Ah, poor Sophie had to endure five-and-a- half years of poverty. What bitter accusations*

25 Ludwig 1931, 130.
26 Ludwig 1931, 131.

I hurl at myself for having so totally forgotten her. At all events, my trip in Sophie's company from Boltenhagen to Klütz would have made for far more pleasant and more interesting memories than a trip around the world.[27]

One may well ask which of these express Schliemann's true feelings – his answer to Sophie's letter, or these lines to his sister. The latter sound hollow, and contain so much special pleading.

Next day he wrote to his sisters:

In vain do I seek to find comfort in philosophical lectures, in vain I visit theatres and art galleries – I can only think of Sophie and her death. I can only blame myself, I can only weep. Now, in my grief, I could spend enough money to travel ten times around the world with her and worship her as a saint. I am ashamed that I answered her in the way I did. On the 10th of October (five-and-a-half months ago), she sent me a strand of her lovely hair. I answered her that I would take it to America with me and that the strand would become my talisman, and that it would protect me against the dangers of the journey. But simply because I had written this, I casually threw the strand into my suitcase, whereas today it is my most precious memento, and I have placed it in a small gold case studded with diamonds, which I shall wear as the most precious jewel as long as I live.

And on the next day he wrote to his sister Louise:

You are to blame for the fact that in 1852 … I did not marry her.[28] *For if you had praised her instead of pointing out some of her peculiar habits – but even without your praise I would have married her, for I was enchanted by her*[29] *… Sophie is the only one who ever loved me; during her illness Aunt Hager wrote that in her imagination and when still conscious she was grateful to me – until her last breath.*[30]

Sophie Bürger

In the midst of securing his divorce from Ekaterina in the USA Schliemann wrote a rather extraordinary letter to his half-brother Wilhelm, who went by the name of Ernst – namely, on 14 April, 1869:

Our views [yours and mine] respecting matrimony do most decidedly not agree, for I consider that marriage is one of the noblest and most holy institutions if its only bonds are love and virtue, but that it is the most degrading and heaviest of fetters if taken by interested motives.[31]

---

27 Ludwig 1931,132.

28 That is the year in which he married Ekaterina. First, Sophie Hecker and Ekaterina Lyshin. Then Sophie Schliemann and Ekaterina Lyshin. All in 1852. An *annus mirabilis* for Schliemann!

29 Here we see unmistakably once again how incapable he was in making up his mind about women, and the degree to which he was dependent on the judgment of others.

30 Ludwig 1931, 132-133.

31 Elli Lily (ed.), *Schliemann in Indianapolis* (Indianapolis 1961), 29.

During this phase, i.e., when Schliemann also wrote to his father about another Sophie,[32] his thoughts oscillated not a little. He also wrote to his brother Ernst on 27 April, 1869 (also from Indianapolis). This has to be one of the most extraordinary letters he left behind:

> *My dear brother,*[33]
>
> *In reference to my last letter I have a great desire to marry you to our cousin Sophie Bürger in Bolzenburg. She is very pretty, blooming, good, natural and clever, and I frankly confess that she is so full of perfection and virtues that I am quite in love with her and have made proposals to marry her in July next. But in doing so my heart had carried away my head, and on mature reflection I find that I am far too old for her, for she is hardly 20, whereas I am 47.*[34] *But she would make an excellent wife for you, for she is brought up in smaller circumstances and by nature economical, modest and she is most aware that economy and industry are two mighty virtues. If you marry her I give her a dowry of twenty thousand francs, with the condition that she puts the money into your business and besides I give you a credit heretofore. But all this is of course only to be done if, as I have reason to expect, she has agreed to my proposal. Besides, a dowry I only give if you marry my dearest Sophie, and I don't give anything at all if you marry another lady. Please despatch at once your consent to Minchen, to send it to Sophie and ask her if she and her parents would not permit you to pay them a visit in end of June.*
>
> *Never* [in] *your life can you get such a charming creature as Sophie. In wishing you a happy wife I remain Your faithful brother HSchliemann.*[35]

The first thing that strikes one in this letter is that Schliemann is already looking around for another wife while still in the middle of attempting to secure a divorce from Ekaterina. But the most extraordinary feature is what he says about his cousin Sophie.[36]

A little later Sophie Bürger comes into view once more, when Schliemann writes a second letter to his brother – i.e., before he had received a reply from Sophie. This second letter, from Indianapolis, dated 19 May, 1869, is still more extraordinary:

> *My dear brother,*
>
> *With reference to my letter of 27th ult I now inform you with immense joy that my dear Sophie has not even deigned to answer my proposal. In fact she must have thought I wanted to laugh at her, for when she wrote me a hot love letter of 10 pages I replied with contempt, whilst now I took that letter up again, discussed it at great length, laughed at it a good deal, accused her of forgery, for she had sent to*

32 See below.

33 Ernst Schliemann (1841-1899) was involved in the wine trade, initially in Bordeaux, then in Berlin.

34 In a letter to Theokletos Vimpos, written on the previous day, he starts off as follows: "Since I have received the pictures you sent me, under no circumstances will I marry the cousin about whom I wrote to you. If she should accept my proposal, then I shall immediately send her a good dowry to enable her quickly to find another fiancé" (Lily 1961, 31).

35 GL Sch BBB 28, 85 (BBB = copybooks of letters).

36 Sophie Bürger is to be distinguished from Sophie Schliemann discussed above.

Elise an envelope with false post stamps and marks to prove that she has written to me in June of 1866 to Paris at a time when I myself had no idea yet of ever going to reside in the Capital of France. Besides I offended her by requesting she should at once dismiss here present bridegroom since he has not the means to support himself and far less to support a family; and after all that abuse I told her I would marry her as soon as I would be divorced. I was much afraid she had at once accepted my proposal, but by merely returning a bill for 100 I had sent her and by not writing a word to me I now clearly see that she is in great wrath against me. She is a clever good looking goodhearted girl and a good contriver in house-keeping. I was sure you would be happy with her, the more so since she would have brought you a fine fortune. I could not have married her for the sake of one [?] for she is too near a relative to mingle. Had she accepted my proposal and had you then taken her off my hands, I would have sent her at once 4000 R__ to get rid of her. Good bye, good business, good health, good happiness. HSchliemann.[37]

Happily, we possess his brother's reply to Schliemann's letter of 27 April, dated Bordeaux, 19 May, 1869, which contains illuminating information:

Mon très frère!

Je te confirme de la tienne de 27 Avril dans laquelle tu me parles d'un mariage entre ta cousine Sophie Bürger et moi.

Je te remercie de tout mon cour, cher Henry, d'avoir songé à moi et la description que tu fais de ta cousine est vraiment séduisant. Du reste Minchen confirme tout ce que dés de sa personne et de ses qualités, mais elle ajoute aussi que Mlle Bürger est fiancée depuis quelque temps avec à un architecte à Schwerin (M. Krüger) et qu'elle ne consentirait pas à rompre avec son fiancé, auquel elle tient énormément.

Minchen t'a écrit ces jours-ci à ce sujet pour te prier d'abandonner tes projets concernant Sophie Bürger. D'après la conviction de Minchen ainsi que d'Elise, elle aime beaucoup trop son fiancé M. K. pour le quitter. -

Je te réspéte, cher Henry, que se te suis reconnaissant de tes bonne intentions pour ma personne. Comme tu partiras probablement fin courant d'Indianapolis, s'adresse ma lettre aux bons sois de Mssrs. L. von Hoffmann Co. New York. Tout à toi ton frère fidèle E Schliemann.[38]

There is also his letter from Bordeaux, of 3 June, 1869:

Mon cher frère!

Comme tu n'a pas reçu de reponse de Mlle Sophie Bürger, tu n'a plus, je me figure, de penser à cet incident. Je vois que tu as l'intention de te chercher une épouse à Athénes; fais bien attention avante t'engager, cher Henry; mais avec ton expérience de la vie tu sauras bientôt te rendre compte des demoiselles qu'on t'as proposées …

37 GL Sch BBB 28, 85.
38 GL Sch BBB 28, 343.

Tout à toi ton frère fidèle Schliemann.[39]

Both letters provide further illuminating background to Schliemann's next encounter.

Sophia Engastromenos

Schliemann's second marriage, namely to Sophia Engastromenos, is too well known to require extensive discussion here. There are, however, several details that are relevant to our topic. The most significant question is: Given the expanse of his relationship with women, how did Schliemann come to choose Sophia? Here too we see him very hesitant in coming to a decision. Despite his age of 47 ½ years, he still appears to be very naive. Given the character of his life up to this point, it seems strange that he should now turn his gaze to Greece.

In any event, he turns to the then Archbishop of Greece, Theokletos Vimpos.[40] He wrote to him in Athens in a letter dated February, 1869:[41]

> *My dear Friend,*
>
> *I send you herewith copies of my book Ithaque, le Peloponnèse et Troie. Please have them bound, keep a copy for yourself, and give the rest to the university. I am sending you a cheque for over 100 francs on Paris to cover the binding. If there is any spare money give it to the poor of my beloved city of Athens.*
>
> *My dear Friend,*
>
> *I cannot tell you how much I love your city and its inhabitants. On the bones of my mother I swear that my entire inclination and intent is to make my future wife happy. Upon my oath she shall never have grounds to complain. And I shall carry her everywhere in my arms if she is good and loving. Here I always find myself in the company of intelligent and beautiful women, who would so gladly soothe my sorrows and spoil me if they knew that I was thinking of becoming divorced. But, my Friend, the flesh is weak, and I am afraid of falling in love with a French woman and once more become unhappy.*[42]
>
> *I therefore beg that you to include a photograph of a beautiful Greek girl. You can buy one at a shop that specialises in photography. I shall always carry this photograph with me in my briefcase, and thereby protect myself from choosing a woman other than a Greek. But if you send me a photograph of the girl you have chosen for me, so much the better. On your oath, choose for me a woman of the same angelic character as your married sister. She should be poor, but educated. She must have a passion for Homer and for the rebirth of my beloved Greece. Whether she can speak foreign languages is irrelevant. But she needs to be*

---

39 GL Sch BBB 28, 344.

40 Prior to his having been elevated to this position, he and his family had lived in St. Petersburg in the 1850s, and where in 1856 he became Schliemann's tutor in both modern and classical Greek.

41 That is, *before* he had submitted his application for divorce!

42 Schliemann goes on at some length on the dangers posed by French women.

typically Greek, have black hair, and, if possible, be beautiful. My chief criterion is that she have an endearing heart. Perhaps you know an orphan, the daughter of an academic, brought up to be a governess, who possesses the virtues which I have stipulated.

My Friend, I open my heart to you, as to a Father Confessor. I have no one in the world in whom I can confide with the secrets of my soul.

Instead of 100, I am sending you 200 francs. Pay the bookbinder, and give any money left over to the poor.[43]

Meanwhile, Schliemann arrived in New York on 27 March, 1869, and promptly journeyed to Indianapolis. But there his application for a divorce became delayed.

Meanwhile, Theokeltos Vimpos answered Schliemann's letter, and included with it at least two photographs of young ladies.[44] Schliemann sent a reply, from Indianapolis, dated 26 April, 1869:

As I am an old traveller and a good reader of faces, I can tell you right away the character of the two girls from their pictures.[45] *Polyxene Gousti's surname shows that her ancestors were Italian.*[46] *Her age is that which is suitable for my wife, but she is bossy, imperious, domineering, irritable and resentful.*[47] *I imagine that she has acquired all these faults in her unenviable profession as a teacher. But I could be mistaken; perhaps if I saw her face to face, I might discover in her a treasure of all the virtues.*[48] *As for Sophia Engastromenos, she is a splendid woman, affable, compassionate, generous, a good housekeeper, lively and well brought up.*[49] *But, alas, she is too young for a man of 47 years.*[50]

Next day Schliemann wrote a second letter to Vimpos (the original is written in modern Greek):

My Friend, already (I have fallen in love with) Sophia Engestromenos ... so that I swear she is the only woman who shall be my wife. But please do not say anything about this to anybody, not even your sister. For two reasons I don't know yet whether I am in a position to marry: first, I am not sure that I shall get the divorce: second, because of my matrimonial (difficulties) I have had no relations with a woman for 6 years ... And if I am convinced that I am fit, then

43 Ludwig 1931, 146-147.

44 Some critics opt for three photographs, but Schliemann seems to imply that there were only two.

45 This strikes one as a considerable exaggeration. If he were such a great reader of faces, one would not expect him to have hesitated so often in connection with women nor constantly to have sought the help of others in reaching a decision.

46 This already was perhaps the decisive factor, since in his initial letter one of Schliemann's basic criteria was that his next wife had to be Greek. In itself the point is irrelevant. Is it possible that Vimpos included Polyxene's photograph in order to fix Schliemann's attention on Sophia, while at the same time creating the impression that despite this Schliemann had to make a choice?

47 Of course, such attributes of character could not be determined on the basis of a photograph – as Schliemann immediately concedes.

48 His judgment appears to be based on, not the photograph, but on the fact that she was a teacher!

49 No more capable of being established on the basis of a photograph!

50 Lily 1961, 31.

> *I shall not hesitate to go to Athens and talk with Sophia, whom I shall marry if she consents. We are forced to wait until then, because if I am impotent … I should not marry".*[51]

On 18 May, 1869, he wrote to his father:

> *The Archbishop of Greece, my former teacher, sent me the photographs of a number of young Athenian ladies from which to choose; from them, I chose Sophia Engastromenos (…) as the most endearing, and it seems in fact that the Archbishop, before being elevated to higher orders and thought that he would still remain a sinner, intended to marry her. In any event, it is my plan, if all goes well, to travel to Athens in July and marry her, and then both of us come to visit you. Since I am entranced by the Greek language, I think I can be happy only with a Greek. But I shall only go for her if she is enthusiastic over scholarship, for I believe that a young, beautiful girl can only honour and love an old man if she has a passion for scholarship, in which he is far ahead of her. I have ordered 12 copies of Sophie's* [sic] *photograph and shall send you one either today or tomorrow morning …*[52]

In his letter to his half-brother Ernst, cited above, Schliemann also wrote:

> *But after all I cannot any longer live alone, and if I succeed in obtaining my divorce I shall most probably go at the end of July to Athens to take a clever Greek wife, the Archbishop of Greece having selected for me several young ladies who for character and accomplishment are, in his opinion, the 'best of the Greek capital'. From among them I have chosen Sophie Engastromenos [from photographs] and if she pleases me so much in person as she charms me in picture I shall no doubt take her. But I shall take more time to examine her character and accomplishments in order not to be cheated a second time. I have ordered copies being made of her photograph and I shall perhaps be able to add a copy to this letter.*[53]

After considerable delay Schliemann obtained his divorce in Indianapolis on 30 June, 1869. He returned to Paris on 24 July. There he had to take care of some business matters, which claimed his entire attention, and so did not arrive in Athens until August. What transpired there is extraordinary, and in many respects comical and pathetic, not to mention also entertaining. First he had to meet Sophia, and then reach a decision. A first, introductory meeting was not enough, especially if one takes seriously what he says elsewhere. In his letter to his brother of 19 May he claimed that, should Sophia prove to be as good as her photograph, he would none the less 'take more time to examine her character in order not to be cheated a second time'. And elsewhere he was not averse to carrying out "a rigorous assessment at least until spring 1870".[54]

---

51 Lily 1961, 33.
52 Ernst Meyer (ed), *Heinrich Schliemann. Briefwechsel* 1 (Berlin 1953), 149.
53 GL Sch BBB 28 85.
54 GL Sch BBB 28, 182.

After their first meeting, which included the family and distant acquaintances, they met once more, but this time alone. Schliemann was determined to ask her, "Why do you want to marry me?" He was completely taken aback and shocked when she answered forthwith, "Because my parents told me that you were a wealthy man". Schliemann immediately took off for his hotel, and there wrote the following note:

> *I was deeply offended, dear Sophia, that as an educated girl you should write to me as if I were a slave. I am an honourable, honest individual. If we are going to become married, can we excavate together, and both become excited about Homer? I am going to leave for Naples the day after tomorrow, and perhaps we shall never see each other again. Should you ever need a friend in the future, contact your devoted Heinrich Schliemann, Dr. Phil., Place St. Michel 6, Paris.*[55]

Had he been true to his principles, he would, of course, have travelled to Paris immediately, but his emotions triumphed over reason – and so he remained in his hotel room, waiting for an answer from Sophia. And an answer came! Being shrewdly counseled by her family, she wrote:

> *Dear Mr. Henry,*
>
> *I am very sorry that you are leaving. You must not take my answer of this afternoon as meant to offend. I believed that a young girl could not answer otherwise. My family and I would be very pleased if you would visit us again tomorrow.*[56]

Had Schliemann been 'an old traveller and a good reader of faces,' he would obviously have noted that she was not the sole author of the answer. Presumably he did not recall the words he had written to his brother a year previously from the other side of the Atlantic about another Sophie (Bürger):

> *I am quite in love with her and have made proposals to her to marry her in July next. But in doing so, my heart has carried away my head and on mature reflection I find that I am far too old for her, for she is hardly 20, whereas I am 47.*[57]

One wonders whether he may, while in his hotel room, have recalled the words which his brother had written to him in his letter of 3 June, 1869:

> *Je vois que tu as l'intention te chercher une épouse à Athènes; fais bien attention avant d'engager, cher Henry; mais avec ton expérience de la vie tu sauras bientôt te rendre compte des demoiselles qu'on t'a proposées.*[58]

One wonders whether he may not also have remembered what he had written to this same brother on 14 April, 1869:

---

55 Ludwig 1931, 150.

56 Ludwig 1931, 150.

57 GL Sch BBB 28, 51. Sophia Engastromenos was only 17!

58 GL Sch BBB 288, 344.

Our views respecting matrimony do most decidedly not agree, for I consider that marriage is one of the noblest and holy institutions if its only bonds are love and virtue, but that it is most degrading and the heaviest of fetters if taken by interested motives.[59]

It would seem that Schliemann was actually rather desperate, for he accepted Sophia's invitation. For how could he pretend something never happened when Sophia had in fact spoken the unvarnished truth? One also wonders what Sophia's family might have thought, could they have known part of what Schliemann had written to his brother in a letter of 8 August on the eve of his departure from Paris for Athens:

Mon cher frère! A ma arrive ici le 5 cit j'ai trouve ta chère lettre du 30 passé …

… Jusqu'au 15 cit je serai très occupé mais lorsque j'aurai remis toutes mes affaires and parfaite règle alors j'ai peur que la solitude ne me paraisse trop accablant et [?] *que je n'ai nulle à Athènes pour me distraire et pour voir si j'y trouve une femme qui me plait. Mais sois sur que je ne me précipiterai pas aveuglément dans un nouveau mariage et que je réfléchierai au mois jusqu'au printemps avant de me decider. On a and Grèce le grande avantage qu'elles femmes sont très pauvres et les instruites et qu'elles font la chasse aux maris. Nous n'estimons pas beaucoup ce qui se présente à nous facilement, c'est ainsi. Il n'y a pas danger qu'on perde sa tête, au contraire on peut les examiner toutes avec sang-froid et pénétrerles moindres détails de leur caractère sans le moindre risque de s'amouracher. Leur desir d'être mariés n'est excellé que par leur chasteté extrême … HSchliemann.*[60]

The details of the meeting with Sophia and her family next day are not known, but one can easily visualise what transpired. In view of Sophia's reply and everything Schliemann had experienced hitherto, it would have been most pertinent for him to 'take more time to examine her character and accomplishments' – and that perhaps until next spring or beyond. But that was not to be. On the contrary, he simply surged ahead, insisting that their marriage take place as soon as possible. And so, on 24 September, 1869, the wedding duly did take place.[61] And so much too for events on that 8 August, 1869, in Athens!

One should, however, recall what he wrote in his letter to Theokletos Vimpos (cited above), of 26 April, 1869, from Indianapolis – namely, in respect of Polyxene and Sophia:

---

59 Lily 1961, 29. By 'interested motives,' Schliemann presumably means 'ulterior motives'. He was probably thinking of Ekaterina's motives and their own disastrous marriage. Schliemann's motives in marrying Sophia strike one as varied and confused. On the day he obtained his divorce in Indianapolis he wrote to Antonios Amiras (30 June, 1869): "I believe, my friend, that I can offer you no better proof of my Philhellenism than this: O forsake American women, the most beautiful women of the world … and go to the far lands of the East to find a Greek wife" (Lily 1961, 48).

60 GL Sch BBB 28, 182.

61 It is probably scarcely worth mentioning that his children were not present for the event.

I think it would be better for me to marry a young widow of exemplary behaviour, who already knows what marriage means. Furthermore, she would be less voluptuous and sensual, whereas young girls believe that heaven and paradise lie in the fulfilment of their physical desires. And I, my friend, used to be very sensual and sentimental. But my character has completely changed. I live now only in a metaphysical world, and I think of nothing except scholarship. Therefore I want a wife only for companionship. I add that my wife must have a great inclination towards learning,[62] *otherwise she will not be able to love and honour me. Try to find me a wife with a Greek name and a soul impassioned for learning. Also, I beg you to write me who was Polyxene's father, what was his profession; when did her mother and father die; how many years has she taught; and what subjects she teaches, and how old she is; to what does she incline; does she know our ancient language?*

Please write to me in Paris, because I think your reply will not find me still in America.

Henry Schliemann, Place St. Michel 6, Paris.[63]

But already on the next day, in another letter to Vimpos, he expressed himself in similar terms in connection with Sophia and her family:

Naturally I do not want a wife with a dowry. On the contrary, if I am physically fit, and propose marriage to Sophia in July, then I shall buy her underclothes and stockings. Nevertheless, I beg you to answer the following questions.

1) Who is Engastromenos? What is his property? How old is he and how many children has he?
2) How many male and how many female? What is the age of each?
3) Particularly, how old is Sophia?
4) What colour is Sophia's hair?
5) Where does the family live in Athens?
6) Does Sophia play the piano?
7) Does she speak any foreign languages? Which?[64]
8) Is she a good housekeeper?
9) Does she understand Homer and our other ancient authors? Or is she really ignorant of our ancestors' language?[65]
10) Would she agree to change her residence to Paris, and to accompany her husband on his journeys to Italy, Egypt and elsewhere? [66]

I warmly beseech you to give me precise answers to all these questions.[67]

---

62 As a teacher, Polyxene was in this respect presumably much more suitable than Sophia.

63 Lily 1961, 31-32.

64 Schliemann was to discover that she did not speak a single foreign language – certainly a major disappointment! In contrast, Polyxene as a teacher was almost certainly capable of speaking several.

65 Here too Polyxene may well have been ahead of Sophia.

66 In this respect, Schliemann may well have concluded that he had virtually another 'Ekaterina' – for although Sophia was to accompany him on various journeys abroad, in the early days of the marriage at least she suffered enormous homesickness for her native Greece.

67 Lily 1961, 33-34.

Soon, however, he suffered from his old vice of uncertainty, as so often in the past. In a further letter to Vimpos from Indianapolis, dated 1 July, 1869, he wrote:

> *Thus I hope on September 4 to be in the fatherland of the gods and heroes. From this country I shall most certainly take a wife, but I don't know whom – Sophia, or Polyxene, or some other".*[68]

Thus at the age of 47 we see Schliemann still in a complete dither over reaching a decision about women. And so it had been all along. At the same time it shows a disturbing immaturity. He is so easily swept off his feet on beholding a beautiful young lady, only to be thrown into confusion and uncertainty when it came to reaching a decision about one of them.

This is also born out emphatically by the contents of his Diary of 1858 – 1859,[69] specifically upon meeting up with James Ferrier, his wife, and, most particularly their daughter, as they travelled in Egypt and Jerusalem. The direct references to the Ferriers are relatively few, but the indirect references are strewn over so many pages. This Diary of 1858-1859 gives us yet another illuminating window into Schliemann's personality, when he was 36 years of age. It is all of a piece with what we have delineated above.

Here, we shall confine ourselves to only a few references to the Ferriers in Schliemann's Diary of 1858-1859, but the reader will, with very little difficulty, be able to extrapolate the references to the young daughter in the pages of the Diary as printed below.

One of the interesting entries is for the last day of 1858:

> *We are sailing across the Mediterranean Sea … This is one of the most pleasant trips I have ever done. Among the other passengers there are two Canadian American families … We were 14 passengers in the first cabin. Among them is a young lady whom I am very fond of … She comes from Montreal, she is plain, she only speaks English and yet her upbringing is excellent.*[70] *It seems that the aforementioned honourable family's name is James Ferrier.*[71]

Likewise,

> *This morning,*[72] *I accompanied Mr. & Mrs. Ferrier on horseback by way of the Jaffa gate to the Damascus gate … ".*[73]

68 Lily 1961, 49.

69 This Diary, in the Schliemann *Nachlass* in the Gennadius Library in Athens bears the Archive title A3 – *Italy, Egypt and the Orient – 1858-1859.*

70 This 'young lady' came from a very devout Christian family, so it is probably not at all surprising that nowhere does Schliemann appear to give the slightest hint that he was at the time married, and had two young daughters (Natalya, born in1855, and Nadezhda, born in 1858).

71 *GL Sch A3*, 70, 73. The Ferriers had two sons and two daughters. The daughter to whom Schliemann refers here must have been Margaret, who later married a John Torrance.

72 April 25, 1859.

73 *GL Sch A3*,186.

Thus, while the reference here to 'a young lady whom I am very fond of' appears on the surface possibly rather casual, it in fact fits into the text of the Spanish Diary and organically into the overall picture of Schliemann's relationship with women. The Spanish Diary turns out to be an integral part of Schliemann's life, in which we find a naivité, immaturity, indecisiveness and lack of integrity.[74] Indeed, it is part and parcel of Schliemann's career as a whole.

74 This is clearly also evident in his first marriage to Ekaterina Lyshin, as Armin Jähne so ably shows below.

Armin Jähne

Die schwierige Ehe von Ekatherina und Heinrich Schliemann

Figure 2: Jekaterina Petrowna Lyshuna (1826-1896)

Bei den gebildeten Völkern steht die Frau seit langem schon gleichwertig neben dem Mann, nicht aber als Sklavin. (Brief vom 25. Juni 1859).

Vorbemerkung

Jeder, der sich später mit der russischen Familie Heinrich Schliemanns beschäftigte oder sich heute mit ihr befaßt, wußte bzw. weiß, dass Schliemanns Ehe mit seiner Frau Ekatherina, geb. Lyshina letztlich scheiterte. Dieses Wissen beinflusst nolens volens die nachträgliche Sicht auf das Verhältnis der beiden Eheleute zueinander und die Interpretation ihrer Beziehungen, einschließlich der Frage, wer die Schuld am ehelichen Zerwürfnis trägt. Auch der Verfasser der folgenden Ausführungen über Heinrich und Ekatherina Schliemann ist sich dieser Gefahr bewußt und vor ihr nicht gefeit. Deshalb versucht er, vor allem an Hand von brieflichen Zeugnissen, insbesondere den Briefen von EkatherinaSchliemann, die innere Entwicklung dieser Ehe zu erschließen, und zwar vom Moment der gemeinsamen Familiengründung an. Er macht sich auf die Suche nach Unverträglichkeiten, die sich zu scharfen Widersprüchen auswuchsen. Sie wurzelten offenbar in persönlichen Eigenheiten, im Charakter der Ehepartner und gerieten zusehends in den Sog starker kultureller Unterschiede, die als äußere Einflüsse das Leben von Ekatherina und Heinrich Schliemann in tragischer Weise belasteten und es letztendlich zerstörten. Auch

Schliemanns Antikebegeisterung, die Neuorientierung seines Lebensplanes, beides von seiner Frau nicht mehr nachvollziehbar, sollten nicht ohne Folgen bleiben. Einige, dem Autor und den Herausgebern von „The Schliemanns Diaries 4" wichtige Briefe Ekatherinas werden in deutscher Übersetzung aus dem Russischen in den Text aufgenommen und kommentiert.[74]

Heirat

Am 12./24. October 1852 (Julianischer/Gregorianischer Kalender) heiratete der Mecklenburger Heinrich Schliemann, seit 1847 Russischer Staatsbürger, in St. Petersburg die Russin Ekatherina Petrovna Lyshina. Die Ehe wurde auf rechtgläubige (russisch-orthodoxe) Weise in St. Petersburg geschlossen, ein für Schliemann, wie sich später herausstellen sollte, folgenschwerer Umstand. Nicht in irgendeiner Kirche fand die Trauung statt, sondern – passend zum eigenen hohem Selbstwertgefühl und zum sozialen Status der Schwiegereltern – in der St. Isaak-Kathedrale, an einem Seitenaltar.

Diese Kathedrale von gigantischen Ausmaßen war die größte Kirche in St. Petersburg und soll bis zu zwölftausend Menschen Platz geboten haben. Erbaut wurde sie unweit des Winterpalastes nach dem Projekt des Architekturzeichners August Montferrand (1786 – 1858), der aber über wenig Bauerfahrung verfügte, so dass das Projekt mehrfach verändert werden musste. Deshalb zogen sich die Arbeiten an der Kathedrale von 1818 bis 1858 hin. Bauherr war letztlich der Zar selbst, damals Nikolai I. Keine Mittel wurden gescheut (23 Millionen Rubel und etwa 410 Kilogramm Gold), um die Kathedrale in aller Pracht zum Ruhme Russlands und des Zarentums entstehen zu lassen. Bei der Feier der großen kirchlichen Feste in der Isaak-Kathedrale waren in der Regel die Zarin und der Zar mit ihrem Hofstaat anwesend. Insofern hat die Kathedrale als eine Art von Hofkirche zu gelten.[75]

Lies schon der Ort der Trauung soziale Bescheidenheit vermissen, so war auch das Lob der Braut – aus Schliemanns Mund – euphorisch maßlos. Am 31. Dezember 1852 schrieb er als Nachtrag im Amerika-Tagebuch: Seit dem 12. October

74 Der Petersburger Journalist Igor Bogdanov hat die Briefe Ekatherinas an ihren Freund und Mann, soweit sie ihm in der Gennadius Library in Athen zugänglich waren, ediert, mit einem Vorwort und jeden Brief mit Kurzkommentaren versehen, die Datierung und die in den Briefen genannten Personen betreffend. Meine Übersetzungen berücksichtigen sowohl die Originalbriefe (als Kopien), als auch Bogdanovs Edition der Briefe. Bogdanov 1998. (Ne privozi s soboju Gomera… = Bring den Homer nicht mit…). Die Briefe werden auch in der Reihenfolge ihrer Veröffentlichung bei Bogdanov angegeben: Nummer und Seite.

75 Das deutsche, aber bescheidenere Gegenstück bildet der Berliner Dom in Deutschlands Hauptstadt, gleich neben dem Stadtschloß der Hohenzollern gelegen. Die Isaak-Kathedrale war seit dem Baubeginn an der einen oder anderen Stelle immer eingerüstet. Deshalb hieß es, dass, wenn die Gerüste endlich ganz fallen, der Zarenthron stürzen werde. Die letzten Gerüste wurden 1916 abgenommen. Im Jahr darauf dankte der Zar ab. In sowjetischen Zeiten war die Kathedrale ein Museum. Jetzt soll sie der orthodoxen russischen Kirche wieder zur religiös-kultischen Nutzung zurück gegeben werden (Informationen zur Isaak-Kathedrale u.a. von Prof. Viktor Karlov, Moskau).

bin ich der glückliche Gatte von Fräulein Ekatherina Lyshin geworden, einer russischen Dame von hohen körperlichen und geistigen Qualitäten. Nun genieße ich alle Bequemlichkeiten eines häuslichen Lebens.[76]

Noch vor der Heirat, aber wenn der Brief sie erreiche, so teilte es Schliemann seinen Schwestern mit, würde er bereits fünf Tage mit seiner Frau zusammenleben und alles Mögliche tun, um sie glücklich zu machen, „denn sie ist ein sehr braves, einfaches, kluges und verträgliches Mädchen u[nd] ich liebe u[nd] achte sie mit jedem Tage mehr". Es folgt dann die Beschreibung der großbürgerlichen, prachtvoll eingerichteten Wohnung in St. Petersburg.[77]

Man mag dem frisch vermählten Ehemann das Lob seiner Gattin nicht verübeln, aber es liegt, scheint es, von Anfang an ein Ton der Falschheit, von etwas Zwanghaften über diesen Schliemannschen Äußerungen. War es seinerseits wirklich eine Liebesheirat oder eher eine Zweckehe? Spricht aus ihm der verliebte Mann, der er angibt zu sein und der sich nun endlich am Ziel seiner Wünsche weiß, oder vielmehr der verstandesgesteuerte Ehemann, der künftige Konsument bürgerlicher Häuslichkeit?

Vieles ist merkwürdig an dieser Ehe von Heinrich Schliemann und Ekatherina Lyshina. Schliemann hatte sich noch vor seiner Amerikareise 1850 bis 1852 um die Russin aus gutem Hause bemüht und ihr den Hof gemacht. Die Lyshins waren eine altansässige, mäßig reiche Juristenfamilie von einigem gesellschaftlichen Einfluss in St. Petersburg. Ihr wurden Verbindungen bis zum Zarenhof nachgesagt. Die Vorfahren stammten aus Russlands Norden, aus der Region Cholmogory, in der auch Michail Lomonosov (1711 – 1765) seine Wurzeln hatte. Sie besaßen ein Gut am Finnischen Meerbusen, in Wammelsu, ein für Russland typischer Ort geselliger, geistiger Begegnungen von Schriftstellern, Malern (u.a. Ilja Repin, 1844 – 1930) und Wissenschaftlern (u.a. der Neurologe und Psychologe Wladimir Bechterev, 1857 – 1927).[78] Mit seiner Heirat trat Schliemann in engere Beziehungen auch zur St. Petersburger Familie Latkin, die mit den Lyshins verwandt und ebenfalls gesellschaftlich bedeutsam war.[79]

Die damals 26 Jahre alte Ekatherina besaß eine überdurchschnittliche Bildung, hatte die deutsche Schule – die renommierte Petrischule – besucht, sprach französisch, spielte Klavier und war sehr selbstbewußt. In den Augen Wilhelmines, der Schwester Schliemanns, war Ekatherina eine Frau von „hoher Geistes- und Herzensbildung".[80]

In solche gutbürgerlichen Kreise einzuheiraten, muß für den jungen Mecklenburger Deutsch-Russen sehr verlockend gewesen sein. Zwar besaß er seit 1847 ein eigenes Handelshaus in Russlands Hauptstadt und war auch im gleichen Jahr in Narva in die dortige zweite Kaufmannsgilde aufgenommen worden, aber seine Vermögensverhältnisse sind mit nachweislich 20000 Silberrubeln Grundkapital als relativ bescheiden einzustufen. Irgendwie schienen ihm auch Zweifel an der Heirat mit einer Russin gekommen zu sein. Im November 1849 wandte er sich

76 Zitiert nach Stoll 1986, 126. Siehe für den originalen Text *The Schliemann Diaries 2*, 189.

77 Brief vom 09. October 1852 aus St. Petersburg, GL BBB V 23/9 (zitiert nach Bölke 2015.187).

78 Andrusová-Vlčeková 1999. 42f.

79 Bogdanov.2002.

80 Brief vom 28. September 1852 aus Patzig, GL B B7 F7/7077. Zitiert nach Bölke 2015. 183.

deshalb „vertrauensvoll" an seine Schwestern um Rat. Er sei sich nicht sicher, ob die vergnügungssüchtigen Hauptstädterinnen auch gute Hausfrauen sein könnten. Vielleicht wäre es besser, eine Mecklenburgerin zu ehelichen.

Wilhelmine, die inzwischen Post von Ekatherina erhalten hatte, antwortete dem Bruder, dass sie keinen Rat erteilen möchte, um nicht an seinem Unglück schuld zu sein. „Mein lieber Heinrich", schrieb sie, „überlege, prüfe und handle selbst, vertraue nicht blindlings Anderer Rat". Und weiter hieß es, er solle „das Vermögen nicht zur einzigen Bedingung bei seiner Wahl machen" und nicht nach Äußerlichkeiten urteilen. Was sie dann hinzufügte, ist höchst bemerkenswert, ja beunruhigend. „Einen Rat will ich Dir noch geben, lieber Heinrich, suche Dich bei den Damen nicht anders zu geben, wie Du wirklich bist, auch versprich ihnen nicht mehr, wie Du nachher halten kannst. – Dein Wesen ist, soviel ich Dich noch von früher her kenne, <u>ernst</u> [so im Original – AJ.], also [ver]suche nun nicht, um den jungen Mädchen zu gefallen, diesen lachenden, heiteren Sinn zu erzwingen – denn man merkt das Wahre doch gleich durch".[81]

Dem Vater gegenüber muß Schliemann, so Wilfried Bölke, generelle Zweifel an Sinn und Nutzem einer Ehe geäußert haben. Der wiederum befürwortete grundsätzlich die Ehe, die nicht ohne Kinder zu sein hat, warnte aber zugleich vor einer übereilten Heirat.[82]

Ekatherina Lyshina zeigte sich ebenfalls skeptisch, ob eine Ehe mit diesem ein wenig sonderbaren, sehr egozentrischen und nicht gerade vermögenden Deutschen das Richtige für sie sei. Jedenfalls gab sie Schliemann, wie es im Deutschen heißt, „einen Korb". Sein Werben um sie hatte ihm keinen Erfolg gebracht.

Im Dezember 1850 machte sich Schliemann auf die Suche nach den Spuren seines in Kalifornien verschollenen Bruders Ludwig. Dort stieg er erfolgreich ins Goldgeschäft ein, handelte mit Immobilien, spekulierte mit Krediten und vermehrte sein Vermögen beträchtlich. Nach anfäglichen Zögern kehrte er im Juli 1852 (nach dem julianischen Kalender) nach St. Petersburg zurück. Hier nahm er seine alte Handelstätigkeit wieder auf und erweiterte seinen Geschäftsbereich, indem er eine Filiale für den Engrosverkauf von Indigo in Moskau eröffnete. Am 24. Juli 1852 (alten Stils) fügte er seinem Amerika-Tagebuch folgende Notiz hinzu: „Heute war ich auf der Börse, wo alle meine Freunde mich mit Begeisterung empfingen. Auch Fräulein Ekatherina Lyshin besuchte ich, der ich in früheren Jahren vergeblich den Hof gemacht hatte, aber jetzt empfing sie mich sehr freundlich, und alles scheint Glück zu versprechen".[83]

Offenbar hatte Schliemann die Hoffnung auf eine Ehe mit Ekatherina Lyshina nicht aufgegeben, zumal er jetzt selbstbewußter und mit einem größeren Vermögen aus Amerika zurückgekehrt war. Angesichts dieser Tatsachen – allein das Sichbehaupten in Amerika verdiente Anerkennung – gab Ekatherina ihre Vorbehalte gegen Schliemann auf und willigte in die Heirat mit ihm ein. Ihr Sinneswechsel, und das ist eine der Merkwürdigkeiten dieser Ehe, wurde

81 Brief Wilhelmines vom 25. November 1849 aus Ramelow, GL B B 5 F 8/5452. Zitiert nach Bölke 2015. 146f.

82 Bölke 2015. 147.

83 Stoll 1986. 126. Siehe den originalen Text *The Schliemann Diaries 2*, 189.

zumindest von Prokofij I. Ponomarev, einem guten Bekannten Schliemanns unter den Petersburger Handelsleuten, mit Skepsis aufgenommen. Wenn sie ihn wirklich gern hatte, fragte er, warum änderte sie ihre Haltung erst, als Schliemann, mit Reichtum gesegnet, seine Geschäftstätigkeit als Großkaufmann auf eine neue, aussichtsreiche Grundlage stellte.[84] Andere werden wahrscheinlich ähnlich gedacht haben wie Ponomarev und nicht weniger überrascht gewesen sein als Schliemanns Vater, der um seinen väterlichen Segen gebeten und formell zur Hochzeit eingeladen worden war.[85]

Offenbar hatten beide Ehepartner aneinander Gefallen gefunden und fühlten sich zueinander hingezogen. Das läßt sich wohl kaum bestreiten. Andererseits scheint ihr Lebensbund keine Heirat aus Leidenschaft gewesen zu sein. Vielmehr dürften sachliche Bewegründe den Ausschlag gegeben haben. Die 26jährige Ekatherina war nach den damals in Russland, und nicht nur in Russland, herrschenden Vorstellungen bereits ein „altes Mädchen" und Schliemann mit seinen über dreißig Jahren ebenfalls nicht mehr der Jüngste.

Dass beide sich nach einem geordneten, ruhigen Familienleben sehnten, ist nur allzu verständlich. Für Ekatherina Lyshina (und ihre Familie) bedeutete Schliemann die Gewähr für materielle Geborgenheit und die Bewahrung des ihrer Herkunft entsprechenden bürgerlichen Lebensstandards und sozialen Status. Für Schliemann brachte die Ehe ohne Zweifel einen sozialen Aufstieg mit sich, dazu die Integration in das Netzwerk der in der Petersburger Gesellschaft fest etablierten Großfamilie der Lyshins und die damit verbundene Akzeptanz in der russischen bürgerlichen Welt überhaupt. Schliemann wußte zudem genau, dass eine solche Liaison seinen geschäftlichen Aktivitäten nur förderlich sein konnte. Insofern ist – über subjektive Gefühlslagen hinaus – eine zweckdienliche gemeinsame Interessenlage von Schliemann und Ekatherina Lyshina, als beide sich zu Heirat entschlossen, durchaus erkennbar.

Briefe aus den Jahren 1852 bis 1856

Es gibt aus dem Jahre 1852 elf kurze, meist undatierte briefliche Äußerungen Ekatherinas vor der Hochzeit. Sie sind vom Inhalt her belanglos. Die Rede ist von Kleiderproben, irgendwelchen Einkäufen, einem Geschenk Schliemanns, von einem Mädchen, offensichtlich einer künftigen, von Schliemann ausgewählten Haushaltshilfe, die Ekatherina aber selbst begutachten will, und von Einladungen ins Haus der Lyshins oder zu ihren Verwandten und Bekannten, die der Bräutigam aber nicht wahrnahm. Seltsam, dass Schliemann diesen vorhochzeitlichen Einladungen meist aus dem Wege ging. In der letzten kurzen Nachricht, schreibt ihm Ekatherina, dass sie es nicht schicklich findet, ihn des öfteren – vor der Hochzeit – in seiner Wohnung zu besuchen.

---

84 Siehe W. Bölke 2015. 173. (Bölke verwechselte die Initialen – nicht I. P. sondern P. I Ponomarev); Mai 1997. 59. Siehe für die Englischübersetzung dieses Briefes *The Schliemann Diaries 2*, 80.

85 In Wahrheit konnte ihm die Reise nach St. Petersburg nicht zugemutet werden. Bölke 2015.181.

Im Januar 1853 war Schliemann in Geschäften in Moskau unterwegs. Im Brief vom 5. Februar dorthin berichtet Ekatherina, dass sie den ganzen Tag über am Klavier übte. Dann kommt eine bemerkenswerte Stelle: 11 Uhr abends fügt sie dem Brief hinzu:

> *Gerade bin ich mit Papachen von Latkins nach Hause gekommen. Dort war fast keiner da und es war ziemlich langweilig. Von Elikonida Nikiforovna [einer Verwandten – AJ.] hörte ich, dass Du ihr gesagt hast, ich würde Dich nicht lieben. Es ist sündhaft von Dir, so etwas zu sagen. Im Leben hat Dich noch niemand so geliebt wie ich. … Wenn morgen keine Briefe kommen, werde ich Dir erst am Sonnabend wieder schreiben. Deine Dich vielliebende Frau Deine K. Schliemann.*[86]

(1) Geradezu entlarvend ist der zweite, schlecht erhaltene Brief vom 21. August aus Neudorf (Novaja Derevnja), einem Datschenort bei St. Petersburg. Zwar fehlt die Angabe des Jahres, aber inhaltlich gehört er in das Jahr 1853, denn Ekatherina schreibt, dass beide einander seit zwei Jahren kennen. Schon die Anrede verwundert.[87]

> *H*[einrich] *Schliemann, meinem Mann. Obwohl schon 21 Tage vergangen sind, seit Du aus St. Petersburg abgereist bist und mich Kranke allein gelassen hast, hast Du Dich die ganze Zeit über nicht einmal bei fremden Leuten danach erkundigt, wie es um meine Gesundheit bestellt ist. Und das ist die Liebe, über die Du so viel quatschst [tolkuesch', abwertend gebraucht – AJ.]! Bis jetzt ist wenig von Deiner Liebe zu spüren, und ich sehe jetzt, dass mir auf dieser Welt kein Glück von dieser Liebe beschieden sein wird, denn wir beide, jeder von uns, verstehen unter Liebe etwas völlig anderes. Ich schreibe Dir all das, um Dir zum letzten Male meine Gedanken auszudrücken, obwohl ich weiß, dass es vergebliche Mühe ist. Du begreifst all das nicht und verstehst es nicht, und deshalb werde ich mich künftig allein um meine Ruhe kümmern. Du sagst mir oft, dass Du mich liebst… Im Verlaufe zweier Jahre, die wir uns kennen, habe ich es nicht geschafft, Dein Vertrauen zu gewinnen. Du blickst auf mich, wie auf eine Verschwenderin und nicht wie auf einen Freund, dem sich der Gatte gerne mitteilt und dem er mit Vergnügen eine Freude bereitet.*

In der hier verdorbenen Stelle geht es um zu wenig Geld, das Schliemann seiner Frau zur Verfügung stellte.

> *Meiner Meinung nach ist es letztlich niederträchtig, Livreen und ähnliche Dinge durch den Lakaien bestellen zu lassen. Ich mische mich da nicht ein, sie können [die Dinge] nach eigenem Geschmack bestellen. Denkst Du wirklich, auf diese Weise Liebe zu erlangen? Selbst ein sparsames Wirtschaften gelingt Dir nicht. In Deinem ganzen Verhalten ist soviel Pedanterie, leeres dummes Zeug. Du guckst immer nur auf die Kopeke und verlierst dabei den Rubel aus den Augen. Das ist eine Milchmädchenrechnung, und in diesem Falle bist Du ein kleinlicher*

86 Brief vom 05. Februar 1853 aus St. Petersburg, GL B 8, 1853, 56. Bogdanov 2002. Nr. 12. 47.

87 Brief vom 21. August 1853 aus Novaja Derevnja, GL B 46, 1861, 764; GL B 37, 1858, 507. Bogdanov 2002. Nr.13. 47f.

Händler und bedauernswerter Mensch. Die besten Minuten im Leben werden Dir unbekannt bleiben. Um mit Dir glücklich zu sein, muß man auf jedem Schritt listenreich und schlau sein.

An dieser Stelle ist von einem schlechten, beleidigenden Scherz Schliemanns und der Angst Ekatherinas die Rede, mit ihm ins Ausland fahren zu müssen.

Wenn ich Dir gesagt habe, dass mir das von Dir zugewiesene Geld nicht langte, dann tat ich aus der festen Gewißheit heraus, dass es nicht gereicht hat. Ich bin jetzt gezwungen bei fremdem Leuten Schulden zu machen, obwohl ich sparsam war und während Deiner Abwesenheit keinen selbst zu einer Tasse Tee eingeladen habe, aber ich war krank, lag eine ganze Woche in der Stadt danieder und mußte viel [Geld] *für Arzneien ausgeben. Verzeih, aus meinem Munde wirst Du künftig weder Beschuldigungen, noch Rechtfertigungen, noch Erklärungen hören. Die Zeit bringt alles an den Tag, und einmal erfährst Du das, was durch nichts rückgängig gemacht werden kann. Ich von meiner Seite werde versuchen, mich so einzurichten, damit Deine Pfeile, die Du auf mich abschießt, auf ein steinernes Herz treffen, das nichts mehr fürchtet. Auf Deinen Wunsch hin wohne ich weiter auf der Datsche und bleibe dort bis zu Deiner Ankunft noch ein/zwei Tage. Danach wünsche ich, in die Stadt zurück zu kehren, weil es, erstens, in unserer Datsche schon kalt, feucht und dunkel ist, und außerdem wäre es kein Wunder, wenn Du mit mir auf Deine Art verfahren würdest. Ich würde mich beleidigt fühlen und etwas Kränkendes sagen, und Du würdest wieder nicht auf die Datsche kommen, und unsere Nachbarn sehen das alles und denken sich weiß Gott was, stellen mir Fragen, die mich in eine unangenehme Situation bringen, so dass ich lieber in die Stadt fahren würde. Lebe wohl, Deine Frau K. Schliemann.*

(2) Der nächste Brief ist ebenfalls undatiert. Gleich zu Beginn findet sich der Hinweis auf Ekatherinas Schwangerschaft, die Schliemann nicht glaubt, weshalb er seine Frau der Lüge bezichtigt. Da der Sohn Sergej am 16. September 1855 geboren wurde, muß der Brief Anfang 1855 geschrieben worden sein. Er ist nicht unterschrieben. Die Anrede lautet „Gnädiger Herr Andrej Aristovitsch".[88]

... Ich bin von meinen armen Eltern gleichwohl in Gottesfurcht erzogen worden und habe bisher niemanden betrogen. Das Schicksal hat es gefügt, dass ich Deine Frau wurde, und, obwohl ich nicht in Leidenschaft entflammt war, als ich Dich heiratete, hatte ich dennoch die feste Absicht, Dir Glück zu bringen. Aber schon von den ersten Tagen unserer Ehe an sah ich, dass ich es mit einem solchen Egoisten zu tun habe, wovon ich mir bislang keinen Begriff machen konnte. Ebenso habe ich bei Dir auch nicht einen Hauch von Takt bemerkt. Darauf hast Du mir noch heute entgegnet, dass ich Dir nicht widersprechen solle, doch dieser Forderung deinerseits fehlt einfach jede Vernunft, denn schließlich bin ich ein lebendiges Wesen und es ist seltsam, von einer erwachsenen Person die widerspruchslose Unterwerfung zu fordern. So auch beim letzten Mal, sage mir bitte, warum es zwischen uns zu unangenehmen Szenen kam. Du wolltest mir an Stelle von 130 Rubeln nur 75 geben. Du meintest, es sei ein Scherz gewesen, aber derartige Scherze kränken einen, und ich habe Dir nicht einmal gezeigt, dass ich gekränkt war. Ich

88 Brief von Anfang 1855, GL B 13, 1854, 806. Bogdanov 2002. Nr. 16. 50f.

bin nur bei meiner Meinung geblieben, dass dieses Geld zu wenig ist, worauf Du so wütend wurdest, dass Du nicht zum Mittagessen nach Hause kommen und außerdem über Nacht wegbleiben wolltest. Du verstehst vielleicht nicht die ganze Gemeinheit dieses Verhaltens, denn sonst hättest Du Dich nicht so benommen. Um Deinen Zorn zu befriedigen, riskierst Du meine Ehre [meinen Ruf], denn was sollen unsere Nachbarn in einem solchen Falle denken. Sie müssen denken, dass ich Dich durch unanständiges Benehmen dazu bringe, sich so zu verhalten. Aber egal, ich schreibe diesen Brief, obwohl ich nicht weiß warum, ich bin traurig und habe niemanden, mit dem ich reden kann. Du wirst wahrscheinlich all das nicht verstehen, denn unglücklicherweise sind wir zwei Wesen, die, wie es scheint, einander niemals verstehen werden. Ich habe das bereits bei unserer ersten Begegnung begriffen und in der letzten Minute meines Lebens werde ich offenbar genauso denken …

(3) Der dritte aufschlußreiche Brief in dieser Reihe mit dem Datum vom 21. April aus St. Petersburg gehört eindeutig in das Jahr 1856, denn Sergej, vom dem auch die Rede in diesem Brief ist, ist noch ein Kleinkind von eineinhalb Jahren.[89] Außerdem korrespondiert der Brief mit Intentionen Schliemanns, die eben in diese Zeit fallen. Im Juni 1856 schrieb Schliemann an seinen griechischen Briefpartner und im Januar 1857 an seinen Pensionsvater Karl E. Laue (1790 – 1860) in Neustrelitz,[90] „dass die Leidenschaft für Wissenschaften so groß bei mir" ist, „dass ich fest entschlossen bin … den Rest meines Lebens meinem Lieblingsfache, den Wissenschaften zu widmen", wobei der Begriff Wissenschaft aus Schliemanns Munde recht diffus klingt.[91] Diese plötzliche Liebe zu den Wissenschaften hatte wohl ihren Grund im Umgang mit Petersburger Gelehrten, die sich zum geselligen Kreis im Hause Schliemanns trafen. Im Briefe Ekatherinas ist diese „Begeisterung" Schliemanns für die Wissenschaften erstmals faßbar.[92]

Mein lieber Freund, Du befindest Dich wohl gerade auf dem Wege nach Rostow und lernst Griechisch, weil Du nichts zu tun hast. So begleiten unsere Gedanken immer die uns nahen Menschen, genau so wie Du wahrscheinlich mehrmals daran gedacht hast, was Kathinka und Serjoscha so machen. Er ißt gerade seinen Kascha (Brei). Die Nanne beteuert, dass er überall den Papa sucht. Diese zwei Tage war ich ständig zu Hause und allein, nur heute früh bin ich zu Latkins gefahren. Morgen will ich in die Kirche gehen. Ich habe viel über unser Gespräch nachgedacht, das wir am Morgen Deiner Abreise geführt haben. Je mehr ich nachdenke, desto mehr gelange ich zu der Überzeugung, dass ein Mensch in Deinen Jahren und mit Deiner Tätigkeit nicht ohne eine Beschäftigung zu leben vermag. Reisen, das Erlernen von Sprachen und selbst Astronomie können nicht

---

89 Brief vom 21. April 1856 aus St. Petersburg, GL B 45, 1861, 388. Bogdanov 2002. Nr. 17. 51f.

90 Bölke 1996. 63f. Anm. 1.

91 Brief vom 11. Juni 1856 an Rhodokanakis (unbekannt), in: Meyer 1953. 82f. („den Rest meines Lebens den Wissenschaften widmen"); auch im Brief vom 18. Juli 1856 an Philipp Kalkmann, in: Meyer 1953. 85; weiter im Brief vom 31. Dezember 1856 an Magdalena Schliemann, in: Meyer 1953. 86f.; Brief Schliemanns an Laue vom 15. Januar 1857 (als Auszug), in: Meyer 1953. 312f. Anm. 120; auch Brief vom 02. April 1858 an Wilhelm Hepner (meine Leidenschaft für Wissenschaften), in: Meyer 1953. 93.

92 Dazu Jähne 2016a. 91f.

Deine ausschließlichen Beschäftigungen sein. Das ist alles gut in Zeiten der Entspannung [in Stunden der Muße] als Vergnügen [Zerstreuung]. Mir scheint, dass Du Dich sehr täuschst, wenn Du glaubst, dass Du in der Lage bist, Dich wie ein Gelehrter mit Wissenschaft zu befassen. Daran muss man von früh an gewöhnt sein. Mir gefällt es, wie Du mit mir am Freitag gesprochen hast. In Deinen Worten steckte viel Wahrheit. Wirklich, eine unabhängige Stellung kann viel Freude (Vergnügen) bereiten, umso mehr jetzt, da wir Serjoscha haben, und etwas zu riskieren, wäre äußerst unvernünftig. Du sagst, dass Du Dich nicht mit kleinen Geschäften abgeben kannst, aber unser Verdienst besteht gerade darin, dass wir unsere Leidenschaften [es folgt in Klammern ein schlecht leserliches deutsches Wort] steuern können. Man muss arbeiten und sich erholen. Diesem Gesetz der Natur begegnet man auf Schritt und Tritt. Leid dem, der es verletzt. Ich habe Dir, wie es scheint, eine ganze Dissertation geschrieben. Leb wohl mein Freund. Ich eile, um Serjoscha zu baden. Ich wünsche Dir Erfolg bei Deinen Geschäften. Nochmals, leb wohl.

Deine Dich liebende Frau K. Schliemann

Diese drei Briefe sprechen auch ohne Kommentare in drastischer Weise für sich. Sie machen dreierlei deutlich: Erstens, steckte in dieser Familie von Anfang an der Wurm gegenseitiger Unverträglichkeit; zweitens, zeichnete sich ein Kulturgefälle innerhalb der Familie ab und, drittens, gab es für Ekatherina als der Hausfrau und Mutter ein ständiges Finanzproblem.

Schliemann begann, nicht lange nach der Hochzeit, seine Frau zu vernachlässigen. Das geschah auf unterschiedliche Weise. Seine Geschäftsreisen waren selbstverständlich unaufschiebbar. Seine Frau mußte sich damit abfinden, und sie tat es auch. Schliemann jedoch verhält sich wenig rücksichtsvoll. Die kranke Ekatherina überließ er sich selbst. Er sorgte weder für eine zeitweilige Hilfe durch eine Krankenwärterin, was in solchen Familien wie der Schliemanns durchaus üblich war, noch erkundigte er sich nach dem jeweiligen Gesundheitszustand seiner Frau. Er schrieb bei längerer Abwesenheit nur unregelmäßig und war an einem engen brieflichen Kontakt wenig interessiert. Schliemann erwies sich als ein ausgeprochener Egoist, gleichgültig seiner Frau und ihren Gefühlen gegenüber. Er verletzte Ekatherina sehr, als er ihre Andeutung einer Schwangerschaft für irreführend hielt und Ekatherina der Lüge bezichtigte, was von mangelnder Sensibilität zeugte. Für Ekatherina war damit – nach reichlich zwei Jahren Ehe – das Maß des Ertragbaren überschritten. Sie reagierte entsprechend und zog in der Familie entsprechende Grenzlininien. Schliemann mußte zur Kenntnis, nehmen, dass er mit seiner Frau nicht umspringen konnte, wie er wollte. Sie trat ihm in den Briefen vom 21. August 1853 (sic!) und von Anfang 1855 als selbstbestimmte und selbstbewußte Persönlichkeit entgegen: keine sklavische Unterordnung unter seinen Willen, keine kritiklose Akzeptanz seiner Wünsche. In einem ähnlich lautenden Brief vom 25. Juni 1859 schrieb sie: „Bei den gebildeten Völkern steht die Frau seit langem schon gleichwertig neben dem Mann, nicht aber als Sklavin".[93]

93 Brief vom 25. Juni 1859 aus Izhora, GL B 38, 1859, 24. Bogdanov 2002. Nr. 31. 64 – 66.

Das sich abzeichnende Kulturgefälle zwischen der gebildeten Ekatherina und dem weitgehend ungebildeten, wenngleich sehr erfolgreichen Kaufmann sollte sich im Laufe der Zeit noch vergrößeren. Im dritten hier angeführten Brief tritt es deutlich zu Tage. Schliemann, und das ist ihm zugute zu halten, war sich seines Bildungsdefizits bewußt, litt darunter und bemühte sich ernsthaft, diesem ihn unbefriedigenden Zustand abzuhelfen. Dazu dienten ihm seine Reisen, seine Sprachstudien und offenbar seine Beschäftigung mit Astronomie, wovon wir hier zum ersten Male erfahren. Ekatherina scheint dafür Verständnis gehabt zu haben – als Freizeitbeschäftigung, als Bildungsvergnügen neben seiner kaufmännischen Geschäftstätigkeit als der eigentlichen Berufung. Zugleich machte sie ihm deutlich, dass sein plötzlich erwachtes Interesse an der Wissenschaft, an einem Gelehrtendasein nur oberfächlich sei und ohne Erfolg bleiben würde, nicht nur weil Schliemann dazu die Voraussetzungen fehlten, sondern weil er von früh an hätte daran gewöhnt sein müssen. Sie sollte mit dieser Einschätzung, schaut man auf Schliemanns gesamtes Leben, nur zum Teil recht behalten. Immerhin wissen wir, wie naiv, unerfahren und großmäulig er seine ersten Grabungen in Troia kommentierte. Schliemann glaubte offenbar, dass mit dem Erlernen des Lateinischen und des Altgriechischen für ihn der einfachste gemeinsame Nenner elementarer Bildung gefunden sei, der ihn in den Augen anderer erhöhen und Anerkennung verschaffen würde, war doch die Kenntnis beider Sprachen in Europa ein Grundbestandteil klassisch-humanistischer Bildung. Auch seine betont herausgestellte Homervertrautheit, sein ständiges „Herumreiten“ auf Homer sollten helfen, das sozial-bedingte Bildungsmanko nach außen zu verschleiern.[94] Einmal bat Ekatherina ihren Mann gar, Homer nicht mitzubringen, „damit wir mit Dir mehr spazieren gehen können“.[95]

Ganz hart tritt im Brief von Anfang 1855 und dem Brief vom 21. April 1856 (nach dreieinhalb Jahren Ehe) und auch in vielen späteren Briefen ein für Schliemann beschämender Charakterzug in den Vordergrund – seiner Ekatherina gegenüber fast grausame Knauserei in Gelddingen. Ein solches Verhalten ist deshalb so schäbig, weil er und somit auch die Familie über ausreichende Finanzen verfügten.[96] Willkürlich setzte Schliemann eine bestimmte Geldsumme fest, mit der Ekatherina während seiner Abwesenheit die eigenen Ausgaben und später den Lebensunterhalt der Familie zu bestreiten hatte. Nun war die Praxis des von der Frau abzurechnenden Haushaltsgeldes damals allgemein üblich (und ist es heute mancherorts immer noch) und an sich nichts Besonderes.[97] Verwerflich bei Schliemann war, dass er Ekatherina von vornherein eine solch karg bemessene Summe zur Verfügung stellte, die natürlich nicht ausreichen konnte, vor allem dann nicht, wenn Krankheit eintrat, Ärzte und Arzneien bezahlt werden mussten oder einfach Sonderausgaben erforderlich wurden. Der einzige Ausweg, der ihr

94 Jähne 2016b. 91f.

95 Brief undatiert, gehört offenbar ins Jahr 1858 (im Brief ist nur von Sergej die Rede).

96 Schliemann hatte sich geschäftlich in Narva etabliert, wo er sich 1853 in die dortige 2. Kaufmannsgilde und 1854 und 1859 in die erste Gilde einschrieb. Die Aufnahme in die erste Gilde setzte ein Vermögen von mehr als 10 -50000, in die zweite Gilde von 5 -10000 Rubeln voraus; Mai 1999. 89 – 94.

97 Den Hinweis auf diesen von mir fast übersehenen Umstand verdanke ich Herrn Manfred Reitz (Dresden/Oberwartha).

blieb, war, sich Geld zu leihen oder Schulden zu machen. Für sie, die Gattin eines überaus erfolgreichen und bekannten Großhändlers, muss diese finanzielle Eingeschränktkeit besonders erniedrigend und für Außenstehende, Verwandte wie Bekannte, unbegreiflich gewesen sein. Einmal wollte er die Summe von 130 Rubeln auf 75 Rubel kürzen. Ekatherina protestierte, machte ihm heftige Vorwürfe. Daraufhin blieb es bei der erstgenannten Summe. Ein anderes Mal (Brief vom 10. Juli 1856)[98] beabsichtigte er, ihr die Tage vom quasi „Kostgeld" abziehen, an denen er nicht zu Haus gegessen hat! Die Klagen um das zu geringe Haushaltsgeld ziehen sich wie ein nicht endender roter Faden durch die meisten Briefe, die Ekatherina schrieb, ein für sie „ewiges" Thema.

Trotz aller dieser innerfamiliären Querelen, scheint mit den Jahren im Hause Schliemann ein wenig Ruhe eingekehrt zu sein, wenigstens zeitweise. Die beiden Ehepartner reden miteinander, auch über Schliemanns Geschäfte. Ekatherina meinte, er könnte seine Geschäfte doch in einem bescheideneren Umfang führen, um weniger zu riskieren. Auch von kleineren Handelsgeschäften ließe es sich gut leben. Diese ihre Ansichten stimmten offenbar mit Schliemanns Vorstellungen überein, denn er habe, wie Ekatherina anmerkt, ebenfalls kein Verlangen nach großem Reichtum. Das verwundert, schreibt er doch in Briefen an die Geschwister, den Vater und an andere Personen, vor allem jedoch in seiner „Selbstbiographie" von der Besessenheit, im Handel möglichst große Gewinne erzielen zu wollen. Am 31. Dezember 1856 äußerte er, rückblickend auf seine Geschäfte während des Krimkrieges (1853 – 1856): „Ich habe hier [St. Petersburg – AJ.] allein einen Kassenumsatz von reichlich 13 Millionen Taler per annum, und in meinem Etablissement in Moskau, welches auch immer mehr wächst, wurden auch reichlich 2 Millionen Taler per annum umgesetzt. Bis jetzt gab Gott viel Glück – allein in 1855 hatte ich einen reinen Verdienst von 220 Tausend Taler".[99]

Andererseits, sagt Schliemann in seiner Selbstbiographie von sich: „Ich bin als Kaufmann ungemein vorsichtig".[100] Er war nicht nur vorsichtig, sondern lebte in ständiger Angst, er könnte sich bei seinen Geschäften verheben und Bankrott machen. Sich diesem Dauerstress zu entziehen, war ein verständlicher Wunsch Schliemanns. Aber auch das wußte Schliemann: Aus Reichtum, aus Geldbesitz fließen Ruhe, Sicherheit und Überlegenheit, öffnen sich Freiräume persönlicher Selbstverwirklichung.

Ekatherina äußerte mehrfach Verständis für Schliemanns Geschäftstätigkeit suchte nach Gemeinsamkeiten mit ihm. In diesem Sinne ist der Brief vom 24. Mai 1859 (aus der Sommerfrische im Dorf Izhora) bemerkenswert. „Gebe es Gott", schreibt sie, „dass wir nur in Frieden und Eintracht leben. Das verlängert unser Leben, verschont uns von Krankheiten und festigt das Glück unserer Kinder; und worüber uns streiten. Ich von meiner Seite hoffe, keinen Anlaß dafür zu geben, indem ich nur für die Kinder lebe und denke, aus ihnen nützliche und gute Menschen zu machen". Sie glaubte, dass Schliemann seinerseits der gleichen

98 Brief vom 10. Juli 1856, GL B, 46, 1861, 643. Bogdanov 2002. Nr. 19. 53f.
99 Mai. 1997. 60.
100 Schliemann 1939. 31.

Meinung ist. „Das Leben vergeht so schnell, und deshalb denke ich, sollte man ernsthaft darüber nachsinnen, die Zeit vernünftig zu nutzen".[101] Dieser Wunsch Ekatherinas sollte nie in Erfüllung gehen.

Briefe vom September 1859

(4) Brief Ekatherina Schliemanns vom 01. September 1859, St. Petersburg[102]

Mein lieber Freund Andrej Aristowitsch,

wir sind schon in die Stadt zurückgekehrt. Ich bin sofort zu Sterkes gefahren, um ihnen darüber zu erzählen. Sie haben mir geraten auf keinen Fall in die neue Wohnung zu ziehen, weil diese, wie sie sagen, feucht ist. Du kannst Dir vorstellen, wie sehr mir an der Gesundheit der Kinder gelegen ist, und deshalb, obwohl ich die Wohnung auf Zuraten von Bruni genommen habe, der sich mir gegenüber verbürgte, dass es nicht die geringste Feuchtigkeit gäbe, bin ich noch heute zu den Bonstedts gefahren. Letzterer sagte mir, ohne selbst dieses Haus gesehen zu haben, er sagte mir, dass die Wohnung in diesem Winter auf jeden Fall feucht sein wird und dass dies überaus gefährlich für die Gesundheit der Kinder sei und ich von dem Gedanken, in diesem Jahr dort zu leben, Abstand nehmen sollte. Gleichzeitig gab er mir den guten Rat, den ich mich beeile, Dir mitzuteilen: diese Wohnung trotzdem per Vertrag für wenigstens vier Jahre zu mieten, sie aber im ersten Jahr mit Zustimmung des Hauseigentümers von Mai bis August abzugeben [offenbar zum Trockenwohnen – AJ.] – bis zum Auslaufen des jetzigen Vertrages. Das heißt wir sollen sie für 100 Rubel billiger abgeben [? – AJ.], aber wir verlieren bereits die 75 Rubel Anzahlung, weil wir diese Wohung jetzt nicht nehmen. Um so mehr, da Du beabsichtigst, noch länger in Paris zu bleiben, und dann ist die Wohnung für uns einfach zu groß. Bonstedt sagt, dass es Zeit kostet, eine Wohnung zu suchen und die Wohnung für uns passend und auch nicht zu teuer zu sein hat. Eine solche Wohnung in der 1. Linie kostet 1200 Rubel und sogar mehr. Diese Wohnung kostet 850 Rubel ohne Feuerholz, aber Holz bekommt man nicht für 150 Rubel. Gereinigt und umgestaltet wird sie auf ihre Rechnung. Du kannst Dir vorstellen, wie unangenehm mir das alles ist. Du kannst aus meinen Briefen ersehen, wie ich mich freue, endlich eine bequeme und hübsche Wohnung zu haben mit einer Küche nebenan und einer warmen Toilette, eine unten gelegene Wohnung mit Balkon, einem Garten und einem schönen Kontor für Dich. Aber die Gesundheit der Kinder steht natürlich höher. Ich bin erstaunt, dass Bruni mir nicht die Wahrheit gesagt hat. Ich habe nicht mit dem Hauseigentümer gesprochen, auch mit dem Architekten nicht. Bruni stand mir näher als alle anderen. Bonstedt sagt, dass es besser ist, den Kontrakt auf länger abzuschließen, da die Eigentümer die Miete sowieso jedes Jahr erhöhen. Unsere Wohnung im Hause von Schadomirskij habe ich noch nicht aufgegeben, weil ich sehe, wie viele Interessenten es dafür gibt. Lebe wohl mein Freund, bleibe gesund und gib mir so bald als möglich eine Antwort und äußere Deinen Wunsch bezüglich der Wohnung. Die Kinder [Sergej und Natalja, geb. im Januar – AJ.] sind Gott sei Dank wohlauf. Nochmals, lebe wohl, ich küsse Dich.

---

101 Brief vom 24. Mai 1859 aus Izhora, GL B 38, 1959, 18. Bogdanov 2002. Nr. 30. 62f.

102 Brief vom 01. September 1859 aus St. Petersburg, GL B 38, 1859, 90. Bogdanov 2002. Nr. 35. 70f.

Deine Dich liebende Frau K. Schliemann

P.S. Bonstedt sagt, dass wir, wenn wir den Vertrag auf vier Jahre abschließen, ihn jederzeit kündigen können, denn es gibt schrecklich viele Wohnungssuchende.

(5) Brief Ekatherina Schliemanns vom 05. September 1859, St. Petersburg[103]

Mein lieber Freund Erice,

Deinen lieben Brief vom 31. August aus Stockholm habe ich erhalten. Ich danke Dir viel-, vielmals für Dein Mitgefühl, das Du mir und Serjoscha gegenüber zeigst, und für Deine Sorgen um uns. Schmit [Schmidt ? – AJ.] habe ich bislang nicht getroffen, ungeachtet dessen, dass ich nach ihm ich weiß nicht wie viele Male geschickt habe. Er lebt fast ständig in Strelna [Vorort von St. Petersburg – A.J.] und kommt nur für sehr kurze Zeit hierher. Das letzte Mal, als ich nach ihm schickte, wurde mir gesagt, dass er mich besucht, sobald ihm das möglich sein wird. Doch jetzt habe ich mich entschlossen, am Montag selbst zu ihm zu fahren und ihn zu bitten, zu mir zu kommen, denn ohne dies macht das keinen Sinn. Im Moment sind wir gesundheitlich zufrieden mit der Ausnahme, dass ich schreckliche Zahnschmerzen habe. Vor einigen Tagen hat sich Serjoscha erkältet und zu husten begonnen. Ich habe gleich nach Dr. [Woldemar – AJ.] Heidike geschickt, der Olga Bonstedt behandelt. Er arbeitet in der Kinderklinik und ist, wie es heißt, ein guter Kinderarzt. Er hat mit geraten, Serjoscha daran zu gewöhnen, morgens ein wenig kaltes Wasser zu trinken, um so der bei ihm ziemlich großen Neigung zu Erkältungen vorzubeugen. Heute morgen bin ich selbst zur Post gelaufen, um zu erfahren, ob von Dir Briefe da sind, denn ich habe schon gestern darauf gewartet. Inzwischen hat mir Sterke Deinen Brief gebracht, der uns sehr gefreut hat. Ich habe mit Absicht Serjoscha daraus viel vorgelesen. Ich weiß nicht, ob er viel verstanden hat, aber er hörte aufmerksam zu. Für Dich sind bisher keine Briefe eingegangen, mit Ausnahme der Börsennachrichten, für die ich gezwungen war zu zahlen. Am ersten [des Monats – AJ.] war ich bei Sterkes, die zu mir sehr freundlich und aufmerksam sind. Ich habe mir bei ihnen 175 Rubel geliehen, von denen ich sogleich 10 Rubel für die Ärzte beiseite gelegt habe. Vor zwei Tagen kam irgend ein Mann zu mir und fragte mich, zu welcher Stunde und an welchen Tagen ich zu sprechen sei, wobei er sagte, dass ihn irgendein Ausländer geschickt habe, der aus Stettin hierher gekommen ist. Dieser weiß zwar, dass Du nicht da bist, aber wünscht dennoch, mich zu sehen und mich zu Dir zu befragen. Ich habe ihm ausrichten lassen, dass ich immer nach 11 Uhr bereit bin, ihn zu empfangen, doch bisher ist keiner zu uns gekommen. Ich weiß nicht, wer das sein könnte. Ich teile Dir noch eine Neuigkeit mit. Sophie Spiglasoff [eine Kaufmannsfrau, verwitwet – AJ.] ist als insolvent erklärt worden und einer ihrer Schuldner, ein gewisser Fedorov, will sie ins Gefängnis bringen. Obwohl für Mme Webber die Idee schrecklich ist, ist sie bereit, das Letzte herzugeben [zur Rettung v. S. S. – AJ.], doch Sophie, die einen sehr starken Willen besitzt, ist damit nicht einverstanden und geht lieber ins Gefängnis, um dort alle die Scherereien loszuwerden, die sie

103 Brief vom 05. Sptember 1859 aus St. Petersburg, GL B 38, 1859, 99. Bogdanov 2002. Nr. 36. 71f.

mit dieser Fabrik hatte und die sie dann umgeht, sollte sich diese Angelegenheit fortsetzen. Die arme Frau haben sie total gequält, und sie hatte keine Ruhe mehr, weder tags noch nachts.

Ich freue mich, dass Du guten Mutes bist. Uns ist in hohem Maße langweilig, besonders leer ist es am Abend. Auf dieser Seite [St. Petersburgs – AJ.] sind wir weit weg von allen Bekannten. Ich weiß nicht, ob unsere Wohnung genügend warm sein wird. Schade, dass die Küche so weit ab liegt. Stell Dir vor, dass an dem Tag, als vor Deiner Abreise das letzte Mal Gäste bei uns waren, Aleksej [vermutlich der Kutscher – AJ.] mir eine hübsche Tasse und Deinen Schirm gestohlen hat. Über Lisabet [eine Bedienstete – AJ.] habe ich viele Häßlichkeiten erfahren, übrigens auch die, dass sie ein sehr enges Verhältnis mit Filofej hatte.

Serjoscha schläft im Moment sehr ruhig. Gebe es Gott, dass sein Husten schnell vorüber geht. Sei Dir gewiß, dass ich für meine Gesundheit und die Serjoschas alles tue, was vernünftig ist und mir dafür nichts zu schade ist. Mit der Droschke fahre ich nicht. Leb wohl mein lieber Freund, mein Segen begleitet dich überall. Wenn mein Brief nicht besonders gut geworden ist, dann ist der Zahnschmerz daran schuld. Leb wohl und bleibe gesund. Ich küsse Dich kräftig-kräftig.

Deine Dich vielliebende Frau K. Schliemann.

(6) Brief Ekatherina Schliemanns vom 12. September 1859, St. Petersburg[104]

Mein lieber Freund Andrej Aristowitsch,

ich kann nicht begreifen, warum ich heute keinen Brief von Dir erhalten habe. Kann es denn sein, dass Du nicht jede Woche einen Brief an uns schreibst.

Ich habe heute einen irgendwie unangenehmen Brief von Dir erwartet, denn ich habe einen schlechten Traum gehabt. Morgen schicke ich [jemanden – AJ.] auf die Post, vielleicht liegt dort ein Brief für mich, obwohl Du die Briefe mir nach Hause schicken wolltest. Was uns betrifft, so sind wir, Gott sei Dank, gesund, außer dass Serjoscha ein wenig hustet und Schnupfen hat, aber ich gehe trotzdem mit ihm einige Male am Tage spazieren. Das Wetter ist wunderschön, im Schatten sind es 15 Grad, ein vollkommener Sommer. Vor einigen Tagen war Gottschalk bei uns. Er wollte Deine Bücher sehen, um irgendetwas, Wechsel betreffend, zu erfahren. Sterkes haben nach ihm geschickt, denn Du hast ihnen nicht gesagt, um welche Wechsel es sich handelt. Zum Glück haben sie schnell gefunden, was sie suchten. Kolja [Nikolaj Petrowitsch – Bruder Ekatherinas – AJ] lässt Dich vielmals grüßen. Er kommt zu uns nur sonntags, denn er wohnt auf der Datsche. Mir hat er ein Paar Kanarienvögel geschenkt, die mit ihrem Gesang unsere Wohnung sehr beleben. Mir ist es schade um die Wasil'evskij Insel. Dort war es irgendwie gemütlich und die Luft besser. Unsere Wohnung [offenbar die neue – AJ.] gefällt allen sehr. Nur eine große Unbequemlichkeit gibt es. Zur Küche ist es zu weit, und sie wird zu schnell kalt.

104 Brief vom 12. September 1859 aus St. Petersburg, GL B 38, 1859, 105. Bogdanov 2002. Nr. 37. 72f.

Einmal waren [Vasilij N. – AJ] Latkin [ein ferner Verwandter von K.S. – AJ.] bei uns. Er hat gesagt, dass Schiruchin schon nach Sibirien gefahren ist, aber dass Wanja durch den Weg [dorthin – AJ.] erschöpft und ernstlich krank ist. Verhüte es Gott, dass dieses Kind stirbt, das Leben würde für diese Frau den Sinn verlieren. Mme Weber [Webber – AJ.] ist schrecklich traurig, Sophie gänzlich verschlossen, doch furchtbar ist das, dass sie das Ganze so hinausziehen [schleifen lassen – AJ.]. Ich weiß nicht, ob ich Dir im letzten Brief geschrieben habe, dass Bokling bei mir war, der aus Liverpool gekommen ist. Er wußte, dass Du nicht hier bist, und ich verstehe nicht, was er hier eigentlich wollte. Es heißt, dass er hier [in St. Petersburg – AJ.] Makler werden will. Philipp ist immer noch auf der Datsche, und wir sitzen die ganze Zeit zu Hause. Von den Gästen waren bislang nur Melin und Mmn Haynahts [eine Freundin von K.S. – AJ.] mit Mann am Abend zum Tee bei mir. Jedesmal, wenn ich diese Frau sehe, muss ich ihren Verstand und ihre Kenntnisse bewundern. Melin und Haynahts redeten über Geschichte, und sie machte solche Bemerkungen, aus denen man schließen konnte, dass sie die Geschichte gut kennt. Außerdem, was die Literatur betrifft, so kennt sie jeden Artikel, der erscheint. Selbst in der russischen Literatur ist sie erstaunlich gut bewandert.

Was mich sehr erstaunt ist, dass es keine Briefe von Dir gibt. Adressiere sie bitte direkt an mich, und schreibe wenigstens einmal in der Woche. Ich vergaß Dir das letzte Mal zu schreiben, dass Lau's [Laue – AJ.] zu meinem Geburtstag hier waren und mir einen hübschen Blumentopf mitbrachten. Ich war damals bettlägerig. Trotzdem bat ich sie zu mir herein. Ich fühle mich jetzt besser. Schmit kommt in den nächsten Tagen von der Datsche zurück, und dann gehe ich ihn besuchen. Leb wohl mein lieber Freund Andrej Aristowitsch. Ich wünsche dir jegliches Wohlergehen auf der Reise und wisse, wenn Du genug von Deinen Reisereien hast und müde wirst, dass Du ein Heim hast und Dir nahestehende Menschen, die sich freuen, Dich wiederzusehen. Deine Dich vielliebende Frau K. Schliemann

Ich schicke Dir diesen Brief nach Stettin, obwohl Du mir in dem Brief, den Du zurückgelassen hast, gesagt hast, ich soll bis 14. September die Briefe nur nach Mecklenburg schicken [H.S. hielt sich angeblich im September 1859 in Mecklenburg auf – AJ.], nun ist morgen der 14. und Sonntag, und es geht keine Post ab. Grüße all die Deinen, und der Schwester Luise wünsche viel Glück.

(7) Brief Ekatherina Schliemanns vom 14. September 1859, St. Petersburg[105]

Mein lieber Freund Andrej Aristowitsch,

Deinen Brief aus Madrid vom 1./13. September habe ich am 12. dieses Monats [alten Stils – A.J.] erhalten. Nun sind schon mehr als drei Wochen seit Deinem letzten Brief vergangen. Du schreibst mir, dass Du nicht einen einzigen Brief von mir erhalten hast, obwohl ich Dir schon dreimal geschrieben habe, jetzt zum vierten Mal. Aller Wahrscheinlichkeit nach wirst Du die ersten beiden in Paris vorfinden, denn Umnov, dem ich meine Briefe auf der Datsche aushändigte, damit er sie an Sterke übergebe, hat sie selbst zur Post gebracht, und ich habe sie

105 Brief vom 14. September 1859 aus St. Petersburg, GL B 38, 1959, 106. Bogdanov 2002. Nr. 38. 74f.

nach Paris poste restante adressiert. Ich bin sehr wütend auf Umnov, dass er so gar nicht entgegenkommend ist und mit fremden Briefen so achtungslos umgeht. Er ist überhaupt äußerst ungefällig. Als wir uns von der Datsche aufmachten, bat ich ihn, uns eine große geschlossene Kutsche zu schicken. Er aber schickte eine offene Kalesche. Das Wetter war kalt, und Serjoscha hat sich erkältet, denn er fühlte sich nicht wohl, als es zurück in die Stadt ging. Mit seiner eigenen Familie würde er nicht so umgegangen sein. In der Stadt angekommen, erfuhr ich von Sterkes, dass er ihnen die Briefe nicht übergeben hatte. Aus Deinen Briefen ersehe ich, das Du Dich amüsierst. Heute war Wasilij N. Latkin bei mir, der sagte, dass die Sache mit Solowjoff ausgezeichnet verläuft [es geht um den Rechtsstreit Schliemann-Solowjoff-Schliemann, 1857 – 1863, der schließlich zugunsten Schliemanns entschieden wurde – AJ.]. Er [Latkin- AJ.] gibt wunderschöne Essen. Unlängst gab es eine großartige Illumination [am 8. September – AJ.]. Franzosen aus Paris haben sie gestaltet. Ich war nicht dort, diese Dinge interessieren mich kaum. An der Polizei-Brücke [Policejskij most] sind viele ertrunken. Verletzte gab es noch mehr. Am Donnerstag ist der Namenstag von Vera [Umnova] und es wird bei ihnen ein großes abendliches Fest geben. Aber dafür muss man als Minimum 25 Rubel ausgeben. Doch dazu bin ich nicht in der Lage. Überhaupt sind 179 Rubel im Monat zu wenig für mich mit den zwei Kindern, und ich kann mir keinerlei Vergnügungen leisten. Jetzt muss ich für Natascha ein Bett kaufen, das mit Matratze 30 Rubel kostet. Außerdem ist noch ein Teppich für sie vonnöten. Sei so lieb und schicke diese Summe für Natascha. Gestern hat mich unser Lakai verlassen. Ihn hat ein adliger Herr zu sich genommen. Es ist sehr schade, denn er trank nicht und war sehr ehrlich. Diesmal hat sich der Mann [H. S. – AJ.] nicht an den Geburtstag seiner Frau erinnert. Du hast ihn sicherlich ganz vergessen, denn in Deinem Brief erwähnst Du ihn nicht. Das ist für mich sehr traurig. Die Kinder sind, Gott sei Dank, gesund. Am 16. dieses Monats wird Serjoscha vier Jahre alt. Ich weiß nicht, ob ich Dir geschrieben habe, dass nach Deiner Abreise Vit bei uns auf der Datscha war und nach Dir fragte. Deinen Brief aus Madrid habe ich Melin vorgelesen, und für kein Geld der Welt möchte ich bei den furchtbaren Stierkämpfen dabei sein, aber das Essen bei der Königin muss sehr interessant gewesen sein.

Leb wohl, bleib gesund. Deine Frau K. Schliemann

(8) Brief Ekatherina Schliemanns vom 15. September 1859, St. Petersburg[106]

Mein lieber Mann Andrej Aristowitsch,

Kaum hatte ich es geschafft, meinen Brief an Dich nach Mecklenburg abzusenden, da erhielt ich Deinen Brief vom 7. September, in dem Du schreibst, dass Du noch nicht einen Brief von mir erhalten hast und ich Dich vergessen habe. Es hat mich sehr bekümmert und erstaunt, denn das ist der vierte Brief nach Deiner Abreise. Den ersten habe ich am 30. August abgeschickt und alle Briefe an die mir von Dir da gelassenen Adressen gesandt. Ich schreibe Dir wöchentlich am Sonnabend, und wenn es passiert, dass ich nicht gesund bin oder nicht schreiben kann, dann werde ich Melin oder Kolja bitten, wenn sie von der Datsche zurück kommen. Bleibe

106 Brief vom 15. September 1859 aus St. Petersburg, GL B 38, 1859, 107. Bogdanov 2002. Nr. 39. 75f.

ruhig und mache Dir keine Sorgen um uns. Heute morgen war Melin bei uns und wollte zum Mittagessen bleiben, aber ich habe ihn nicht eingeladen, denn ich hatte nichts vorbereitet. Es langte allein für uns. Ich muss ziemlich eingeschränkt wirtschaften und „mich zur Decke strecken". Unsere Wohnung verlangt, dass ich sie sauber halte. Auch die Ärzte sind zu bezahlen. Schmit, so heißt es, darf man nicht weniger als 10 Rubel geben. Dann kommt er zweimal. Gedike habe ich gleichfalls 6 Rubel geschickt. Ich lasse nicht anschreiben, sondern zahle sofort. Auf diese Weise weiß ich immer, was ich [an Geld zur Verfügung – AJ.] habe. Ich hoffe, dass Du bei Deiner Rückkehr mit Deiner Frau zufrieden sein wirst. Serjoscha hat sich sehr über Deine Karte aus Stockholm gefreut. Wir sind, Gott sei Dank, gesund. Kolja läßt Dich vielmals grüßen.

Deine Dich vielliebende Frau K. Schliemann.

Meinen Brief vom 30. August habe ich durch Filofej abschicken lassen. Hat er mich etwa betrogen? Diesen Brief schicke ich am 16. ab, denn mir ist gesagt worden, dass montags keine Dampfer fahren. Heute hat Serjoscha Geburtstag, und seine ersten Worte waren, als er aufwachte: „Papa ist zum Geburtstag nicht gekommen, doch Weihnachten wird er da sein". Wir haben keine Gäste eingeladen. Wenn jemand von selbst kommt, dann ist es gut. Nochmals leb wohl, mein lieber Erice, und schreibe öfter.

(9) Brief Ekatherina Schliemanns vom 20. September 1859, St. Petersburg[107]

Mein lieber Mann Andrej Aristowitsch,

es ist seltsam , dass ich schon wieder keinen Brief von Dir erhalten habe. Schreibe mir bitte wöchentlich, denn wenn ich solange keine Nachrichten erhalte, dann denke ich, dass Du entweder nicht gesund oder ärgerlich auf mich bist. Übrigens, so scheint es mir, habe ich nichts Unrechtes getan. Wir leben völlig zurückgezogen; an Sergejs Geburtstag hat uns niemand besucht. Er ist ein ganz liebes Kind, ganz von der Art, wie ich Kinder liebe. Er schwatzt im Moment sehr viel, mitunter ganz vernünftig. Doch ist er ein schrecklicher Wildfang und so hat er sich gestern zum Beispiel Salz in die Nase getan und dann lange geweint. Vor drei Tagen sind wir mit der Kutsche zu William [Ritter ? – AJ.] gefahren. Sie waren sehr liebenswürdig und gaben uns auf, Dich vielmals zu grüssen. Auch bei Umnowychs sind wir gewesen. Sie haben im Hause von Heimburger keine Wohnung gefunden und sind deshalb bei Solodownikow eingezogen [beide sind Hausbesitzer in St. Petersburg – AJ.]. Aber diese Wohnung ist nicht billiger als die andere, 1500 Rubel wie es scheint und 300 Rubel für Umbauten, und eine andere Treppe müssen sie selber besorgen. Bei Heimburger sind Freriks, Voelkel, Schtriter, Totleben und Flug eingezogen.

Schreibe mir bitte mehr. Dein letzter Brief war sehr kurz, mit dem ersten hingegen war ich sehr zufrieden. Er war ganz einfach und ich erkannte Dich darin wieder, und das scheint mir einer der ersten Vorzüge eines Briefes zu sein. Le styl c'est l'homme, so sagte es, scheint mir, Voltaire. Ebenso finde ich in diesem Brief eine wahrhaftigere Anhänglichkeit als in einem der Briefe sonst, die ich von Dir erhalten

107 Brief vom 20. September 1859 aus St. Petersburg, GL B 38, 1959, 111. Bogdanov 2002. Nr. 40. 76f.

habe. Die Liebe zu mir hat Dich, wie es scheint, in diesen 8 Jahren nicht verlassen [ist nicht erkaltet – AJ.]. Heute sind wir mit Serjoscha viel spazieren gegangen. Es herrscht hier ein erstaunlich warmes Wetter, 10 – 12 Grad im Schatten. Heute wollte Melin zu mir zum Mittagessen kommen. Einen besonderen Luxus wird er bei uns nicht finden, Wein wird es nicht geben, dafür habe ich eine Flasche Honig [Honigwein, Met – AJ.] gekauft. Man hat mir schon die Doppelfenster eingesetzt und dafür ordentlich Geld verlangt [etwas viel verlangt – AJ.]. Jetzt fühle ich mich besser. Die niedrige Treppe ist für mich eine große Erleichterung. Nur der kalte Eingang in die Küche ist in dieser Hinsicht unangenehm. Ich tröste mich damit, dass uns keine Wahl blieb. Barteling [Thomas – AJ.] habe ich gestern getroffen. Er lebt noch auf der Datsche. Leb wohl mein lieber Freund Erice, bleibe gesund und fröhlich. Kolja lässt Dich vielmals grüssen. Ich habe ihn eine ganze Woche nicht gesehen. Ich schicke Dir in dieser Woche noch einen zweiten Brief. Nochmals, lebe wohl und schreibe öfter. Deine Dich krepko [kräftig] küssende Frau K. Schliemann

(10) Brief Ekatherina Schliemanns vom 27. September 1859, St. Petersburg[108]

Lieber Andrej Aristowitsch,

man muss „zornig" schreiben oder noch andere Adjektive (deutsche) verwenden, solche, die ich machmal bei unseren Streitereien gebrauche. Ich sage nur, dass Dein Brief aus Stettin mir Verdruss bereitet hat, dass ich viel geweint habe, weswegen ich krank wurde, und mir das Wort gegeben habe, künftig mir nichts mehr so zu Herzen nehmen und besser auf meine Gesundheit zu achten, die mir durch nichts zurückgegeben wird. Aber Dich bitte ich, mit mir vorsichtiger umzugehen. Wenn Du aus meinem Brief vom 31. August [er ist offenbar nicht aufzufinden – AJ.] herausgelesen hast, dass ich neidisch auf das glückliche Zusammentreffen von Umständen bei Vera [Vera Umnowa – AJ.] sei, so weiß ich nicht, was dabei für Dich so beleidigend und ein solches Verbrechen sein könnte. Nehmen wir an, dass Du mir deshalb eine Bemerkung hättest machen können und ich das eingestanden hätte, und damit wäre die Sache erledigt gewesen, aber aus einer Entfernung von Tausenden von Werst einen Brief zu schreiben und ihn nur damit zu füllen, das ist wahrlich zu viel. Bei uns beiden gibt es Unzulänglichkeiten und man kann sich bemühen, sie zu beseitigen, aber nicht derart schroff und grob. Unsere Wohnung ist voller Liebreiz, und Du hast von Deiner Seite getan, was Du tun konntest. Und wenn Du nicht abgereist wärest, dann hätten wir wahrscheinlich eine Wohnung mit einem Zimmer mehr genommen, wo wir mit Dir hätten bleiben können. Wir hätten eine weniger prunkvolle, aber bequemere gefunden, aber jetzt sind die Reden darüber zu Ende. Ich reiche Dir die Hand zur Versöhnung und will keine Minute mehr darüber nachdenken noch reden. Gestern war der Namenstag Sergejs. Die Latkina [eine Verwandte – AJ.] hat uns besucht und ihm ein schönes Spielzeug mitgebracht und deshalb habe ich sie [die Latkins – AJ.] für morgen zum Mittagessen eingeladen.

108 Brief vom 27. September 1859 aus St. Petersburg, GL B 38, 1859, 116. Bogdanov 2002. Nr. 41. 78f.

Gib Dir bitte Mühe, mir zu schreiben, was ich mit den Briefen und den Börsennachrichten machen soll, die hier auf Deinen Namen ankommen. Du hast mir gesagt, dass ich Dir die Börsennachrichten nicht mitteilen soll. Warum soll ich sie dann annehmen? Unlängst brachte man ein ganzes Paket davon und forderte dafür 90 Kopeken in Silber. Ich habe darüber Sterke befragt. Sie sagten mir, dass Beste sei es, sie nicht anzunehmen. Brief habe ich einen erhalten, und seinen Inhalt teile ich Dir mit, soweit ich ihn verstanden habe. Dieser Brief kam aus Manchester von einem Hugo Liepenbruck... Ihr Agent Otto Sitthorn lebt in Petersburg und macht verschiedene Geschäfte in Moskau und Petersburg... Er [H. Liepenbruck] will von Dir wissen, wie es um seine [O. Sitthorn] Moral und Solidität bestellt ist, und außerdem wie man in Petersburg mit Bankrott gegangenen Kreditgebern verfährt. Das ist alles, was er schreibt. Aber ich fürchte, das ich Briefe aus Amerika, besonders in Englisch geschriebene, überhaupt nicht verstehe. Seltsam ist dabei, das auf den Briefen „bezahlt" steht, doch ich musste 33 Kopeken zahlen. Der alte Sterke war auf den Tod krank, jetzt bessert sich sein Zustand. Das Wetter bei uns ist sehr warm. Kolja läßt Dich vielmals grüssen. Er ist noch auf seiner Datsche. Morgen hoffe ich, einen Brief von Dir zu bekommen, einen etwas freundlicheren. Lebe wohl, ich küsse Dich kräftig, Deine Frau K. Schliemann

Im nächsten Briefe schreibe ich ausführlich, was mir Schmit gesagt hat.

Die hier angeführten sieben Briefe, die allesamt im September 1859 geschrieben wurden, enthalten viel Belangloses über das Leben der Familie Schliemann in St. Petersburg, und doch zeigt sich in diesen Alltäglichkeiten, dass es dem Miteinander der beiden Ehepartner an Normalität fehlte. Was bereits oben zu den zitierten und analysierten früheren Briefen Ekatherinas gesagt wurde, wird durch die Briefe vom September 1859 nochmals bestätigt.

Schliemann vernachlässigte seine Frau auch weiterhin. Er läßt sie mit ihren Sorgen, insbesondere dem permanenten Finanzproblem, allein, kümmert sich nicht um ihre Gesundheit und ihr Wohlbefinden, überläßt ihr die offenbar nicht einfache Wohnungssuche bzw. die Regelung der Wohnsituation und schreibt nur unregelmäßig Briefe, oftmals mit ungerechtfertigten Vorwürfen versehen. Bei der Geburt der Tochter Natalja war er nicht anwesend, auch nicht bei ihrer Taufe. Er vergißt den Geburtstag von Ekatherina, und traurig ist die Bemerkung seines Sohnes Sergej: „Papa ist zum Geburtstag nicht gekommen, aber Weihnachten wird er da sein" (das ist er dann auch). Ende Dezember 1858 hatte Schliemann eine Orientreise begonnen. Im Juni kehrte er zurück nach St. Petersburg und brach schon Mitte August zu einer neuen Reise nach Spanien auf, die bis zum 06. October 1859 dauerte. Schliemann war von zwölf Monaten nur knapp vier Monate bei seiner Familie!

Ekatherina ihrerseits ist für jede positive Gefühlsregung Schliemanns dankbar und empfindet sie als Anerkennung für ihre Leistungen im Interesse der Familie. Einmal schreibt sie ihrem Mann als Antwort auf einen seiner Briefe mit Freude, dass seine Liebe zu ihr in den acht Jahren Ehe offenbar nicht erkaltet sei. Das war am 20. September 1859. Sieben Tage später muss sie sich wieder – im Zusammenhang mit der Petersburger Wohnung – gegen seine haltlosen Anschuldigungen wehren. „Wir beide haben unsere Fehler, und man kann sich bemühen, sie zu beseitigen,

Figure 3: Sergej – Natalia & Nadeshda Schliemann

aber nicht derart schroff und heftig“. Sie will ihm die Hand reichen und alle Mißverständnisse sollten vergessen sein.

Ekatherina war kein „Frauchen“, kein Hausmütterchen, sondern eine selbstbewußte, starke Persönlichkeit. Sie wußte um ihren Platz, ihre bedeutsame Rolle in der Familie. Sie war viel zu sehr auf sich allein gestellt, um das nicht zu begreifen. Ekatherina offenbarte sich in ihren Briefen als liebende Frau („Deine Dich vielliebende Frau“, „целую Тебя крепко“ – das schreibt man nicht so einfach dahin), als sich sorgende Mutter, als keineswegs verschwenderische und verantwortungsbewußte Ehepartnerin. Nüchtern stellte sie fest, dass sie sich abends oft sehr einsam fühlt, zurückgezogen lebt und äußere Geselligkeiten auslassen muß, weil ihr die finanziellen Mittel fehlen, z. B. 25 Rubel. Schliemann hingegen amüsierte sich auf seinen Reisen. Wiederholt teilte Ekatherina ihm mit, dass ihn zu Hause Fürsorglichkeit und ein gemütliches Heim erwarten, ihm die Familie, wenn er nur will, sozialer Ruhepol sein kann. Natürlich liebte Ekatherina auch Klatsch und Tratsch, und Schliemann wohl auch.

Aus den Septemberbriefen erfahren wir noch etwas Anderes und durchaus Erfreuliches. Nur im Brief vom 12. September 1859 ist davon die Rede: Schliemanns führten in St. Petersburg – vor allem, wenn er anwesend war – ein gastliches Haus, zumeist am Sonntag. Es trafen sich dort Verwandte, Freunde, Kaufleute (Unternehmer), Vertreter des St. Petersburger Kultur- und Geisteslebens.[109] Der Brief ist deshalb bedeutsam, weil er vermuten läßt, dass dieser Kreis interessierter Leute, die bei Schliemanns verkehrten, größer war als wir bisher wissen. Ekatherina schreibt, dass bei ihr zu einer abendlichen Teerunde Heinrich (Andrej Andrejewitsch) Mellin (vielleicht Mellien, 1786 – 1861, AJ.) und Helene (Jelena) Haynahts (eine Freundin von K.S. – AJ.) mit ihrem Mann zu Gast waren. Mellin lehrte Geschichte und Statistik an der Deutschen Hauptschule bei der Kirche der „Heiligen Ekatherina“ in St. Petersburg und war ein guter Bekannter der Schliemanns. Madame Haynahts war eine enge Vertraute Ekatherinas und ihre Gesellschafterin auf den Reisen nach Westeuropa. Ihr Mann Peter arbeitete als

109 Dazu kurz Jähne 2016a. 91f.

Hauptrevisor für eine Versicherungsgesellschaft. „Jedesmal", merkte Ekatherina an, „wenn ich diese Frau sehe, muss ich ihren Verstand und ihre Kenntnnisse bewundern. Mellin und Haynahts redeten über Geschichte, und sie machte solche Bemerkungen, aus denen man schließen konnte, dass sie die Geschichte gut kennt. Außerdem, was die Literatur betrifft, so kennt sie jeden Artikel, der erscheint. Selbst in der russischen Literatur ist sie erstaunlich gut bewandert".

Dresden 1866

Am 10. Juli 1866 trat Schliemann seine so genannte „Wolgareise" an, die ihn ursprünglich ans Kaspische Meer, in den Kaukasus und weiter führen sollte. Die Reisepläne mußten unterwegs geändert werden. Daraufhin ging es den Don abwärts, die Küste der Krim entlang, über Odessa die Donau aufwärts, dann mit der Eisenbahn über Dresden, Leipzig, Nürnberg, Genf nach Paris. Aus dieser Reise wurde, was Schliemann zu diesem Zeitpunkt nicht ahnen konnte, ein endgültiger Abschied von Russland und seiner Petersburger Familie.[110] Die wenigen Tage in Dresden, die Absichten, die Schliemann mit der sächsischen Residenz verband und die von nun an den Briefwechsel der Jahre 1866 bis 1869 mit Ekatherina bestimmten, führten letztendlich zum Bruch in der Familie. Aus Meinungsverschiedenheiten wurden unüberbrückbare Gegensätze, die keinem der Ehepartner mehr Raum zur Versöhnung ließen. Weder Ekatherina noch Schliemann vermochten es, über ihre eigenen Schatten zu springen.

> *Dresden, Montag, 26. September, Hotel Belle Vue* [deutsch]:
>
> *Mein Koffer traf endlich gestern Morgen 2.30 Uhr in Bodenbach [heute Děčin – AJ.] ein, und verließ ich daher letzteres mit dem Zug von 6.45 Uhr und traf um 9 Uhr hier ein. Die Bahn geht immer am Ufer der Elbe entlang durch die Sächsische Schweiz und überall ist die Aussicht prachtvoll und erhaben. Wir fuhren links bei der herrlichen Festung Königstein vorbei, die auf einem 900 Fuß hohen Felsen liegt. Den neben derselben liegenden Wald hat der Commandant der Festung fällen lassen. Da hier die Steine zum Bauen [bei Schliemann, zu Bauten' – AJ.] von den Felsen am Ufer der Elbe geschnitten werden und nur ein Geringes kosten, ist hier die Erbauung von Häusern äußerst billig und läßt sich für 8000 Reichsmark ein großes zweistöckiges Gebäude errichten. Die Stadt [Dresden – AJ.] besteht daher aus großen schönen Häusern, und die Straßen sind gepflastert mit breiten Trottoiren. Ich besuche hier wiederholt das schöne Johannisbad in der Elbe[111] und die Bildergalerie. Ich fuhr heute mit einem Lohndiener fast den ganzen Tag umher, besuchte die Blindenanstalt, wo blinde Mädchen feine Stickereien ausführen und unter anderem Perücken machen und Männer Körbe flechten, Seile machen usw., besah die neuen Bauten als Landhäuser in der Alt- und Neustadt und gefielen mir besonders diejenigen in der Schillerstraße am Ufer der Elbe und ganz besonders das von Herrn Felsner mit Turm, vielen Balcons*

110 Jähne 1999. 101 – 108.

111 Schliemann liebte es, in offenen, kühlen Gewässern zu baden und zu schwimmen. Die schnell fließende Elbe war kalt und damals noch sehr sauber. Der Autor hat als Kind und Jugendlicher in den 1950er Jahren selbst noch in der Elbe gebadet (oberhalb Dresdens).

und großem Garten mit Springbrunnen, welches wahrscheinlich für 35000 Rubel käuflich ist. Besuchte die Pensionats von Madam Pauline Cachin, Dr. Ad. Müller, Frau Dr. Clara Pfotenhauer geb. Lafargue und gefielen mir besonders das letzte sowie die Lehr- und Erziehungsanstalt von Dr. Christ[ian] *Fr[iedrich] Krause.*[112] *Während ich dort war, waren die Kinder gerade unter Musik mit Exerzieren und Turnen beschäftigt unter Aufsicht mehrerer Lehrer. Des Doktors Mitteilungen über die Art und Weise des Unterrichts, die Waschungen der Kinder den ganzen Körper morgens und abends mit kaltem Wasser, die strenge Aufsicht zur Bewahrung vor geheimen Lastern, die Rettung zweier Zöglinge vom Onanismus, die Gewandheit der Primaner im lateinischen Sprechen, die ausgezeichneten Tiermalereien – alles bezauberte mich.*

Dresden, Mittwoch, 28. September [französisch]:[113]

Gestern stattete ich Herrn A. Rothe, Schillerstraße 25, einen Besuch ab. Ich habe hier im Hotel Herrn und Frau Haynahts [?] getroffen, die nach Dresden gekommen sind und hier in der Chemnitzer Straße Nr. 6 ein Haus gemietet haben mit einem ausgedehnten Garten für 250 Rubel im Jahr. Mit dem Makler Ernest Luckner, Ostraallee 12 parterre, der mir von Herrn Kaiser, dem Besitzer unseres Hotels, empfohlen wurde, habe ich mir heute die Villen von Renz, gegenüber von Rothemunde, von Felsner, Gerstkam, Seibt (neben dem Bad), das Hotel Oppenheim in der Stadt, die Gebäude der englischen Botschaft, ebenso der preußischen Botschaft angesehen (sie wäre lukrativ, wenn man ihr Gebäude in Ordnung brächte und das Gelände verkaufte), weiter die Rost-Villa, wo vorne die Adelsons und Plasowitschs wohnen, die Villa von Orland[e] und die von Fohde [oder Tohde?], für die ich 35 oder 30 Tausend Rubel geben würde, aber außer den Villen von Gerstkam und Seibt finde ich keine unter ihnen, die mir zusagen würde.

Dresden gefällt mir sehr, sowohl unter dem Gesichtspunkt seines exzellenten Klimas, als wegen seiner schönen Lage über der Elbe, wegen seiner herrlichen Gärten und Promenaden und wegen seiner grandiosen Galerie von Meisterwerken der Malerei, unter denen sich der Liebesgarten von Rubens befindet, mit den drei Frauen und seiner Hölle, in die von Teufeln alle diejenigen hinab gestoßen werden, die ihm geholfen haben, die Comtesse zu verführen, die danach seine Frau wurde, während er selber von einem Drachen verschlungen wird. Was den Charakter der Bewohnerschaft anbelangt, ist hier alle Welt wahrhaft freimütig und loyal, und man schafft schnell Kontakt zum Fremden und gewährt ihm freigiebig seine Freundschaft, ohne ihn je zu demütigen. Ich bin nie in einer Stadt gewesen, in der man mit größerer Höflichkeit und Achtung behandelt wird. Es scheint mir, dass man sich mit Lust bemüht, wahr und liebenswert zu sei".

An eben diesem Mittwoch, am vorletzten Tag seines Aufenthaltes in Dresden, schrieb Schliemann erneut einen Brief an seine Schwester Doris in Roebel.

---

112 Die Erziehungs- und Bildungsanstalt von Dr. Christian Friedrich Krause in Dresden feierte 1866 ihr 25-jähriges Bestehen. Siehe Hempel 1866; Rachel 1907 (die Literaturangaben verdanke ich Herrn Prof. Siegfried Wollgast, Dresden).

113 Für die Übersetzung aus dem Französischen danke ich Herrn Prof. Hans-Otto Dill, Berlin.

Er beginnt wie folgt:

> *Liebe Schwester seit ich Dir vom Süden Russlands schrieb [gemeint ist der angeblich aus Astrachan abgeschickte Brief vom 18./30. Juli 1866], veranlaßte mich ein Fieber die Reise nach Persien und dem Kaukasus aufzugeben, die Häfen des Asowschen Meeres und Schwarzen Meeres zu besuchen und die Reise über Dresden nach Paris fortzusetzen; von dort denke ich, über Tunis und Constantinopel zurückzukehren [nach Sankt Petersburg – AJ.].*[114] *Viel Interessantes habe ich gesehen, seitdem ich Dir schrieb, und mein damals erschöpfter Vorrat von guter Laune ist jetzt wieder größer als je. Ich bin entzückt von den hiesigen Schulen und beabsichtige ganz hierher zu ziehen, um die Kinder zu erziehen. Meine Frau muss es noch nicht wissen…*[115]

Aus Dresden schickte Schliemann auch einen Brief an seine Frau in St. Petersburg. Ihre Antwort ist nicht datiert. Die Anrede lautet knapp: Andrej Aristowitsch. Das Ende des Briefes hat sich nicht erhalten. Wahrscheinlich unterschrieb Ekatherina mit „Deine Frau K. Schliemann". Es heißt dort:

> *Deinen Brief aus Dresden habe ich vor drei Tagen erhalten, und es ist für mich angenehm zu sehen, dass Du in guter geistiger Verfassung bist. Du schreibst mir auch über die Einrichtung von Dr. Krause, die Dir sehr gefallen hat und die sich tatsächlich einer gewissen Bekanntheit erfreut. Aber ich habe erst unlängst in einer deutschen pädagogischen Zeitschrift gelesen, dass die privaten Lehreinrichtungen in Deutschland sehr auf den Effekt ausgerichtet sind und sich zu sehr den Wünschen der Eltern anpassen.*

Weiter schreibt sie, dass sie für ihre Kinder keine Gouvernanten wünsche. Sergej sei bereits zu groß, und für die Töchter Natalja und Nadeschda könne sie selbst die Rolle der Gouvernante übernehmen.

> *Vor allem bin ich bemüht, ihnen [den Kindern – AJ.] Gottesfurcht beizubringen und sie in dem Bestreben zu fördern, ihre Handlungen dem Gewissen und der Vernunft unterzuordnen.*[116]

Im nächsten Brief teilt Ekatherina ihrem Mann mit, dass sie Sergej eingeschult habe. Er lerne jetzt in einer Schule, die ihr von Bernhard von Dorn (1805 – 1881) vorgeschlagen wurde, einem Bekannten der Familie und Professor an der St. Petersburger Universität, und die Sergej auf das Gymnasium vorbereite.[117]

Schliemann hatte sich, wie im Brief an Schwester Doris angedeutet, in seinem Schreiben zurückgehalten, seine wahren Absichten verschwiegen, also die Katze noch nicht aus dem Sack gelassen. Die Entscheidung jedoch war bereits gefallen. Ohne sich mit Ekatherina zu besprechen, hatte er für sich

---

114 Im Brief vom 30. März 1866 aus St. Petersburg (GL BBB V 26 / 41), also noch vor der so genannten Wolgareise, schrieb Schliemann an seinen Vetter Adolph: „Wahrscheinlich werde ich … über Odessa u(nd) Sevastopol nach Constantinopel reisen, um den Kampfplatz von Troia und Ithaka zu besuchen, auch Athen wieder zu sehen". Nach Bölke 2002. 327. Dieser Reiseplan wurde nicht ausgeführt.

115 Brief vom 27. Mai 1866, an Doris (Dorothea) Petrowsky in Röbel. Meyer 1953. 128.

116 Undatierter Brief aus St. Petersburg, GL B 60, 1866, 377, 3. Bogdanov 2002. Nr. 176. 208.

117 Ebenfalls undatiert aus St. Petersburg, GL B 60, 1866, 377. Bogdanov 2002. Nr. 177. 209.

festgelegt, den Lebensmittelpunkt von St. Petersburg nach Paris oder Dresden zu verlegen, wobei er Dresden wegen seiner Mittellage den Vorzug gab. Der Streit, der daraufhin zwischen den Eheleuten entbrannte und im Winter/Frühling 1866/1867 kulminierte, überstieg alles bisher Dagewesene. Nicht nur das Kulturgefälle in der Familie verstärkte sich. Hinzu trat als äußerer Faktor der für Russland charakteristische Gegensatz zwischen Slawophilen und Westlern, jener Dualismus von Europaphilie und Europaphobie (Europa gleich Westeuropa). Ekatherina weigerte sich strikt, mit dem Kindern nach Paris oder Dresden zu ziehen, obwohl ihr Vater in Dresden gestorben und begraben worden war.[118] Sie versteifte sich auf ihr Russentum, dass sie auch für Ihre Kinder in Anspruch nahm, beharrte auf den Vorzügen des russischen Bildungswesen, dass nicht schlechter sei als in Europa und berief sich auf mentale Unterschiede. Schliemann bittet, bettelt geradezu, droht, heuchelt, versucht es mit Erpressung, schaltet Freunde und Bekannte ein, um Druck auf Ekatherina auszuüben. Er verliert jegliches Maß und jede Kontenance. Er diskreditiert Ekatherina auf eine Weise, die nicht einmal auf den Ruf der eigenen Person Rücksicht nimmt. Gegenüber seinem Vertrauten Baron Konstantin von Fehleisen (1804 – 1870), den Schliemann für sehr diskret hielt, äußerte er:

> *Nach der Geburt des Sohnes, „widerstand sie jede Nacht meinen Annäherungen und schrie, ich würde versuchen, sie umzubringen, und ich kann Ihnen sagen, dass ich seit 1855 an ständig Diebstähle an meiner Frau begehen musste, um ihr meine zwei nachfolgenden Kinder zu stehlen“.[119] Wie sind diese Diebstähle [was immer darunter zu verstehen ist – AJ.] vonstatten gegangen?*

Der ausufernde Streit um die Übersiedelung der Familie nach Westeuropa, nach Dresden oder Paris, der die Emotionen auf beiden Seiten hochkochen ließ und der einer gründlichen Untersuchung bedarf, führte schließlich zum Auseinanderbrechen der russischer Familie Heinrich Schliemanns.[120] Daran konnte auch Schliemann Reise im Januar 1869 nach St. Petersburg nichts mehr ändern. Der episodenhafte Kurzaufenthalt dort endete in einem Desaster. Schliemann mußte das russische Reich fluchtartig verlassen.

Mit Dresden und den Folgen der Schliemannschen Entscheidung hatten beide Seiten ihren auf unterschiedliche Weise geführten Kampf um die Familie verloren.[121]

Im Sommer 1868 hatte Schliemann seine schicksalhafte Begegnung mit Frank Calvert. Das war der eigentliche Wendepunkt in Schliemanns Leben – der Entschluß, Troia auf dem Hügel Hissarlik zu suchen. Im März 1869 wurde er offiziell amerikanischer Staatsbürger; am 30. Juni 1869 erfolgte in Indianapolis (USA) die Scheidung von seiner Frau Ekatherina, ein Rechtsakt, der in Russland ungültig war. Am 24. September 1869 heiratete er die Griechin Sophia Engastromenos

118 G. Andrusová-Vlčeková, a.a.O., S. 43 (der Vater Petr Alexandrowitsch Lyshin).

119 Nach Traill 1995. 31 (für die Nachprüfung dieser Stelle danke ich Herrn Reinhard Witte, Ankershagen/Waren). Brief vom 06. Februar 1867 (BBB 27, 10 – 18).

120 Es geht dabei um die Hintergründe, die Motive, die Art und Weise der Argumentation und um rechtliche Fragen.

121 Dazu auch Bölke 2002. 331 – 340.

(1852 – 1932), und im Frühjahr 1870 führte er eine noch illegale Probegrabung auf dem Hissarlik durch.

Seinen finanziellen Unterhaltsverpflichtungen gegenüber seiner russischen Familie kam Schliemann uneingeschränkt nach und traf für sie Vorsorge über seinen Tod (1890 in Neapel) hinaus.

Da wir wissen dass Schliemann das Erlebte oftmals beschönigte,[122] müssen wir fragen, wie er das Essen der spanischen Königin, wovon Ekatherina in ihrem Brief schreibt, beschrieben hat?

Fazit

Im Leben Heinrich Schliemanns sind zwei strikt von einander zu trennende Perioden zu unterscheiden: erstens, der vortrojanische Schliemann und, zweitens der Troiaausgräber Schliemann. Zwischen beiden Phasen gibt es nur wenig verbindende Linien. Zu unterscheiden sind weiterhin der Geschäftsmann Schliemann, der er ein Leben lang blieb, und der Archäologe Schliemann, der sich allmählich zu einem Fachmann der Spatenwissenschaft entwickelte, und schließlich der Mensch Schliemann.

Als Mensch ist Schliemann umstritten. In seiner russischen Familie zeigte er sich als Egoist, ja als Haustyrann, als ein deutsch verdorbener komischer Ehemann, der alle Klischees bediente, die in Russland seinen Landsleuten traditionell angelastet wurden. Er war in Geldsachen ausgesprochen knausrig bis zur Herzlosigkeit, obwohl er anderen gegenüber durchaus großzügig sein konnte. Er vernachlässigte Frau wie Kinder, war kein Gentleman, sondern vom Charakter her seinem Vater ähnlicher als er es wahrscheinlich selbst wahrhaben wollte, und er blieb ein Leben lang ein Plebejer. Darüber können im Laufe des Leben erworbene Verhaltensweisen und Umgangsformen, auch eine gewisse altersbedingte Abgeklärtheit, Bescheidenheit und Güte nicht hinwegtäuschen.[123]

Zwei harte Köpfe, zwei unterschiedliche Lebensauffassungen, ja Kulturen stießen in der russischen Ehe Schliemanns aufeinander. Das konnte auf Dauer nicht gut gehen.

---

122 Siehe z.B. den Untergang der "Dorothea" bei Texel. Arentzen 2011.
123 Siehe dazu Stucky 2016.

Wout Arentzen

Schliemann in Spain

The extent of Schliemann's interest in politics is unclear. During the Crimean War from 1853 to 1856, he managed to increase his wealth greatly by evading the English naval blockade of Russia. He has described the methods he used to achieve this and the anxiety he suffered in the process in a number of letters. There is however nothing in these letters that reveals his opinion of the war.

> *Es war in der Zeit des Krimkrieges. Da die russischen Häfen blockirt waren, mussten alle für Petersburg bestimmten Waaren nach den preussischen Häfen von Köningsberg und Memel verschifft und von dort zu Lande weiter befördert werden* … [Ich] *machte grosse Geschäfte in Indigo, Farbhölzern und Kriegsmaterialien (Salpeter, Schwefel und Blei), und konnte so, da die Kapitalisten Scheu trugen, sich während des Krimkrieges auf grössere Unternehmungen eizulassen, beträchtliche Gewinne erzielen und im Laufe eines Jahres mein Vermögen mehr als verdoppeln. [It was in the time of the Crimean War. Since the Russian ports were blockaded,, all goods intended for Petersburg had to be shipped to the Prussian ports of Köningsberg and Memel, and transported further from there by land* … [I] *made big deals in indigo, dyeing wood and military materials (saltpetre, sulphur and lead), and since the capitalists were wary of engaging in big enterprises during the Crimean War, I was able to make substantial profits and more than double my capital in the course of a year.*[124]

Since the diary of his travels through Spain includes a prognosis of the political developments that reveals his dislike of England at this stage of his life, it might be thought that as a naturalized Russian this would lead him to support the Russian side in the Crimean War. There is no reason to believe this, however, since he certainly had no evident dislike of the other parties to the conflict, France, the kingdom of Sardinia and the Ottoman Empire.

It is difficult to tell to what extent his political prognosis was influenced by his wish not to offend the Canadian politician James Ferrier (1800-1888), the father of the young woman Schliemann adored. Ferrier was born in Scotland, a country for which Schliemann expressed great admiration in his prognosis.

Apart from this prognosis, Schliemann only made incidental references to the politics of this period. In order to clarify the political situation behind this diary, I will go into this topic in somewhat greater detail. Events in the Ottoman Empire will not be discussed here. They were already dealt with comprehensively in Part III of *The Schliemann Diaries* and play no further role in Schliemann's memoirs here.

---

124 Schliemann 1892, 19-21.

Celebrations in Paris

The Congress of Vienna in 1814-1815, which regulated European affairs after the fall of Napoleon (1769-1821), also changed the face of Italy. Austria received Venice as far as Lombardy and Dalmatia. Genoa, Savoy and Nice were added to the kingdom of Sardinia. This region was ruled by Charles Felix (1765-1831), who was succeeded after his death by Charles Albert (1798-1849). Francis, the son of Archduke Ferdinand of Austria-Este, became Francis IV, Duke of Modena (1779-1846). He was succeeded by his son Francis V (1819-1875). Maria Louisa of Spain (1782-1824) was appointed regent for her underage son Charles Louis in the kingdom of Etruria and the Duchy of Parma. The Papal States regained their secular power. The third son of Charles III of Spain (1716-1788) became Ferdinand I (1751-1825), king of the Two Sicilies. He was succeeded after his death by his son Ferdinand II (1810-1859). England gained Malta, and Monaco and San Marino became independent. The states formed in this way were all subject to authoritarian regimes, which led to protest among their population. Various secret societies were formed, aimed at the unification of Italy. This led to uprisings, for example in Naples and on Sardinia. The political situation became so unstable after 1830 that order could only be maintained with the aid of Austria. Pius IX (1792-1878) was elected as Pope in 1848, against the wishes of Austria. He reorganised the Papal States, which reinforced the call for reform in the other states. Charles Albert responded by granting Sardinia a constitution and promising to support Lombardy in its struggle against Austria. He was seen by many as the liberator of Italy, but his troops were no match for the Austrian army and he was forced to sign a cease-fire with Austria in 1849 after a defeat. He abdicated and was succeeded by his son Victor Emanuel II (1820-1878).

Meanwhile, Louis Napoleon (1808-1873), the nephew of Napoleon Bonaparte, has gained power in France. He was elected President of the Republic in 1848, and after a *coup d'état* he took the throne in 1852 as Napoleon III. He was determined to play a leading role in European politics. He signed a treaty with Sardinia in 1852, according to which Northern Italy would belong to Sardinia while Savoy and Nice would become French. To seal this union, Jérôme Bonaparte (1784-1860), the youngest brother of Napoleon I, married Maria Clothilde (1843-1911) the daughter of Victor Emanuel II.

This treaty and the resulting demands that Napoleon III made on Austria led the Austrian forces to attack Sardinia on 1 January 1859. The French army came to Sardinia's assistance, and proved to be too strong for the Austrians. After a number of bloody battles, a cease-fire was signed in Villafrance on 11 July. Austria lost Lombardy, but Napoleon III did not dare to conquer the whole of Italy as he had planned, because he feared Prussian interference. Though he had not achieved his goal and the peace was only temporary, he celebrated it in great style.

These are the celebrations to which Schliemann refers at the start of the diary dealing with his travels through Spain. The English were suspicious of the French expansionist ambitions, and since the evidence of the diary indicates that

Figure 4: Napoleon III at the Battle of Solferino, 24th of June 1859, – by Adolphe Yvon (1817-1893)

Schliemann had a very poor opinion of England and the English at this time it is worth while investigating what the English thought of these celebrations.

The English government took a cautiously positive view of the French activities in Italy. After all, France and England had been allies in the Crimean War. However, this alliance came under great pressure after the attempt by the Italian Republican Felice Orsini (1819-1858) to assassinate Napoleon III with a bomb. This attack left 8 people dead and 142 wounded, though the Emperor and his wife were unharmed. The discovery that Orsini had used three bombs that had been made in England led to a wave of anti-English feeling in France.

The Italian republicans did all they could to widen the gap between the two great European powers. One of the three Italian statesmen who were ultimately responsible for the unification of Italy, Giuseppe Manzzini (1805-1872), published a letter to the Editor of *The Times* on 10 August 1859 in which he warned the English of the dangers of their friendly attitude towards the French.

> *If Europe persists in the cowardly and guilty indifference with which she now contemplates the progress and the acts of re-born Bonapartism she will inevitably expiate it bitterly.*
>
> *As the expedition of Rome was the prelude to the French coup d'état, the Crimean war and the Italian war mark two steps towards the European coup d'état. The next step will be the dismemberment of Turkey and war against England, isolated in Europe.*

... Had the expedition of Rome been prevented – had Lord Palmerston,[125] *instead of declaring through Lord Normanby*[126] *that the Protestant Government of Her Majesty approved the restoration of Catholicism in Rome, said to Louis Napoleon, "England does not admit foreign intervention in the internal affairs of other States," Imperialism would have been arrested in its commencement ... The weakness of others has alone rendered Louis Napoleon daring and powerful. The expedition of Rome had the triple object of propitiating the French clergy, of discrediting the Republican Assembly, and of preparing the French soldiers to fight against the Republican flag at home. But, besides this, it was for Louis Napoleon a means of ascertaining the tendencies and strength of the Governments of Europe.*[127]

These warnings do not seem to have made much impression on the English, who were full of wonder and expectation at the preparations being made for the entry of troops into Paris.

Paris at the present moment resembles some veteran coquette under the hands of her tire women, making up for a Court ball, at which she is bent on transcending all her former appearances. Unfortunately, the city, unlike the lady, cannot adorn herself in the privacy of a boudoir, thence to emerge rouged, flounced, and befeathered, with every flower and jewel in place, to dazzle all beholders. The Parisians are condemned to witness the laying on of the paint and the twining of the wreaths, to stumble over the undressed bars of the crinoline, and to observe every stage of progress, from nudity to full dress. Just now bare poles and wet plaster are the order of the day.[128]

The correspondent of *The Times* sent an account of the preparations to his paper on 14 August. I include a few fragments of his long, colourful narrative below to give some idea of the festivities. Sitting at a window looking out on the *Boulevard des Italiens*, he wrote the story that would have to be posted the same afternoon. He was not the only one on that street.

Right and left, looking from a window that commands the whole of that Boulevard, the eye is first caught by the multitude of human beings stationed in every place where they can possibly find a footing. The broad asphalt footpaths are the places least densely packed, although a line of spectators five or six deep, many of whom have been there for the last two or three hours, occupies that portion of it nearest the carriage way. But between them and the houses circulation is still possible, although here and there impeded by gathering masses, which threaten soon completely to choke the way ... That the windows and balconies are full need hardly be said; in the more lofty of the latter benches have been erected higher than the top of the balustrade, to enable a second row of spectators to obtain a view of what passes 60 feet below them.[129]

Not only the balconies but the roofs were full of spectators.

---

125 Henry John Temple, 3rd Viscount Palmerston (1784-1865).

126 Constantine Henry Phipps, 1st Marquess of Normandy (1747-1863).

127 Giuseppe Mazzini *The Times* 10 Aug. 1859.

128 *The Times* 13 Aug. 1859.

129 *The Times* 16 Aug. 1859.

He has placed himself at the very extremity of a high wall rising from the roof of a house fully 80 feet above the street. The smallest push – a puff of wind, one would think – would suffice to send him flying through the air, to be dashed to pieces on the pavement, unless he were previously impaled upon the chevaux-de-frise[130] *which stick out from the wall a dozen feet below him. He looks down into the street, however, with the most perfect coolness, and even rocks himself backwards and forwards, as if he thought his position not only airy and elevated, but highly eligible.*[131]

After the public had been waiting for hours, the spectacle finally began.

It is now a quarter past 10, and the head of the great military pageant has halted at the western extremity of the Boulevard Montmartre. A small advanced guard, the very apex of the column, is composed of a detachment of the Cent Gardes. Five gorgeous trumpeters ahead, and then a score of the chosen troops of the brilliant little corps. They halt opposite the Opera Comique. Away to the left, where the Emperor and his staff have made a brief pause, is heard a clamour of applause. Presently the column moves on, the Emperor in front, occasionally taking off his cocked hat, and frequently touching it with the military salute. He is in the undress of a general officer, blue frock and broad red ribbon of the Legion of Honour. He is well received … many cries of "Vive l'Empereur!" are distinguishable through the confused uproar. Bouquets are thrown from the windows to the Emperor and his Staff … [He is followed by] *the wounded, lame and maimed, with arms in slings and pallid visages in the order of their regiments. First are the Chasseurs [light cavalry], then the Voltigeurs [light infantry], then the Zouaves [soldiers of Algerian origin] and the Grenadiers …*[132]

These too are pelted with flowers. The parade continues with one regiment after another. The correspondent begins to fear that he will be unable to describe the crowning point of the parade because his dispatch needs to be posted, but fortunately it arrives just in time.

Here come the Austrian standards, four in number. Two of them are in pretty good condition, and, as the wind blows out their folds, the double eagle is displayed upon the yellow ground. A third is borne in its oilskin case; of the fourth only a few tatters still cling to the battered pole. They are borne by the men who took them, escorted by two soldiers of each regiment of the corps to which each captor belonged. The flags were taken by the Guard, 1st, 2nd, and 4th corps. Then come the guns, easily distinguishable by the make and colour of their carriages. Upwards of 30 go by, and the march continues; more troops, and still more, a continued broad stream glistening with bayonets pouring over the low brow of the Montmartre Boulevard, surging onward and past us and disappearing in the gulf of the Rue de la Paix.[133]

130 A protruding stake used to close a breach in a wall.

131 *The Times* 16 Aug. 1859.

132 *The Times* 16 Aug. 1859.

133 *The Times* 16 Aug. 1859.

The correspondent continued his story on the next day. After a further description of an almost endless series of regiments marching past, he gave the readers some information about Paris itself.

> *The public buildings and a great many private houses were lit up with gas and coloured lanterns, and in many places coloured lamps were hung from the branches of trees, The garden at the back of the Elysée was beautifully illuminated. Strings of coloured lamps were suspended completely across the Boulevard Montmartre. The little garden round the triumphal column at the corner of the Opera Comique was lighted up, and the gleam of the lamps on the shrubs and flowers produced a very pretty effect. A loyal tailor on the Boulevard des Italiens had embroidered his premises with foliage, supporting numerous lanterns of coloured paper, but, by some misfortune, nearly all the candles went out at a very early hour, giving rise to a suspicion that he had been making des économies de bouts de chandelle* [had been penny pinching]. *A line of gas jets ran all round the Place Vendôme. On the Place du Carrousel the triumphal arch was surmounted by numerous coloured lamps, while columns of gas supported eagles, stars, and the letter "N" with a crown above it. Before every window of the Tuileries and of the new Louvre were placed the Imperial arms … The Parisians and the hundreds of thousands of strangers who had flocked hither for the great festival were doubtless tired by their early rising and by the exertions and excitement of the day, for after 11 o'clock the crowd in the streets rapidly thinned, and by 1 they were as deserted as usual at that hour.*[134]

But that was by no means the end of the festivities. They continued the next day, concluding with a fireworks display in the evening.

> *The double festival of the triumphal entry and the Emperor's birthday*[135] *terminated, as it commenced, prosperously, and without a single drawback. The sky, yesterday forenoon menacing, brightened as the day wore on, and again threatened rain before night arrived, but finally cleared; the wind abated, and fireworks and illuminations met with no impediment. The former lasted but a short time, and persons who are judges of such matters do not speak very highly of them, although the concluding bouquets were certainly beautiful. The illuminations were most effective. The garden of the Tuileries was like a scene from fairyland. The tower of St. Jacques de la Boucherie, illuminated to the very summit, was a very striking object. On the south side of the Seine the electric light dispensed its dazzling radiance. Above the Hotel de la Legion d'Honneur the cross of the order was elevated. To all appearance it floated in the air. The contrivance that supported it far above the roof of the building was not visible through the darkness, and one saw only a cross of coloured fire, set in an expanded jet.*[136]

Thus, though no progress at all was being made in the peace negotiations, Paris celebrated Napoleon III's victory.

---

134 *The Times* 17 Aug 1859.
135 Napoleon III was born on 20 April.
136 *The Times* 18 Aug 1859.

The Spanish-Moroccan War of 1859-1860

Spain grew into a world power under the Habsburgs, who ruled from 1504 to 1700. They were succeeded by the Bourbons, whose extremely reactionary style of government under the influence of the highly conservative Catholic clergy put a stop to most forms of progress and caused Spain gradually to lose more and more influence in the world. This trend was intensified when Napoleon occupied Spain in 1807. Charles IV abdicated, and the Bourbons lost control not only of Spain but also of the colonies where local governors and freedom fighters took advantage of the power vacuum to declare independence. In this way, Spain ultimately lost most of its colonies.

Figure 5: Isabella II (1830-1904)

Ferdinand VII (1784-1833) regained the throne in 1814, after the Peninsular Wars concluded in the defeat of the French forces. He used brutal methods in an attempt to rid the country of anyone with liberal sympathies but was unable to prevent conflicts not only between the Conservatives and the Liberals, but also within the ranks of these two movements. After the birth of his daughter Isabella in 1830, Ferdinand's wife Maria Christina persuaded him to set aside the Salic Law which would have prevented Isabella from succeeding him. This set him at odds with his brother Don Carlos (1788-1855) and his followers the Carlists, while the Republicans wanted to do away with the monarchy entirely.

When Ferdinand died in 1833, he was succeeded by the infant Isabella II (1830-1904) who was only three years old at the time. His widow declared herself regent. Carlos opposed this move, claiming the throne for himself as Carlos V. He was supported by the Basque provinces, Navarra and the clergy. The Liberals and moderates who supported the right of Isabella to the throne were backed up by Great Britain, France and Portugal. This conflict led to the first Carlist War, which lasted from 1833 to 1839 and ended in the defeat of the Carlists.

The regency came to an end when Isabella was thirteen, when she was deemed capable of ruling herself. She had a limited education and was very religious, but this did not stop her from leading a very licentious life. She married her cousin Fransisco de Asis de Borbon (1822-1902) but had a string of lovers, including all the ministers who served her. She had 12 children, of whom only four reached adulthood. Most historians are of the opinions that her husband was not the father of any of these children.

In 1859 Spain was in danger of losing its Moroccan possessions of Ceuta and Melilla. After the loss of all its other colonies, it had no wish to see these cities go and it was now in a position to do something about it. It started reinforcing the forts around Ceuta, which had been in its possession since 1688. This displeased Sultan Abd-al-Rahman ibn Hisham of Morocco (1778-1859), who thought it was time his country regained possession of these two cities. He ordered the army and the people of Morocco to attack the Spanish forces. The local tribes followed his orders, and managed to damage the forts.

> *The Gibraltar Chronicle of the 29th* [August] *has the following intelligence from Ceuta:*
>
> *At half past 5 p.m. on the 27th the Moors showed themselves on the heights of what is called Old Ceuta, and made themselves strong behind the walls; but a steamer and a launch were sent in that direction, and opened fire, the garrison at the same time firing some shells, which killed 14 Moors, and compelled them to disperse. On the 28th, at the same hour, they again made their appearance, and the same events took place as on the preceding day. Troops were expected to-day at Ceuta, which, according to intelligence received, must have embarked at Alicante ...*

Intelligence from Algeciras (published in the Gibraltar Chronicle of the 5th [September]*) states that two battalions of light infantry had arrived there in the short space of three days, by rail and steam, via Alicante, and were, for the present, billeted upon the inhabitants. Other troops were shortly expected, which, when they arrived, would be encamped between Algeciras and San Roque.*

The master of a boat which arrived at Gibraltar on the morning of the 5th from Ceuta reported that there had been no firing between the garrison and the Moors on the 2nd and 3rd. A body of about 1,200 men sallied out on the afternoon of Saturday to make a reconnaissance, and went as far as the place called "La Mezquita" (the Mosque), without meeting with any Moors. The troops destroyed all the parapets or obstructions in their way, and then returned to the fortress without firing a shot. On Sunday afternoon the Moors made their appearance again in front of Ceuta and began to annoy the garrison, which fired several rounds of grape and solid shot at them.[137]

It is clear that the local tribes were no match for the Spaniards, and the Moroccan government was doing its best to put the genie back in the bottle.

A rumour has been current here [Tangier] *for some days past, to the effect that Sidi Mohamed El Katib*[138] *... was using his utmost endeavours to appease the Kabyles in the vicinity of Ceuta, reminding them that they were committing a great sin in opposing the will of the Emperor; for, according to the precepts of the Mohammedan religion the Sultan is entitled to a blind obedience on their part, as he is considered to be inspired by the divine spirit, and acts under the express commands of the Almighty; but, in spite of this, it seems that the other precept, which says 'that he who slays an infidel in the war will gain paradise', has rather more influence over them.*[139]

In order to intensify the struggle, the Moroccan authorities had declared the battle for the two cities a holy war against the infidel, while the Spaniards made a similar effort to give the conflict a religious dimension.

The fortress of Ceuta now represents the brilliant und useful past [of Spain]. *That consideration sufficiently explains the feeling which impels Spain to preserve entire ... that Christian civilization...*[140]

Catholic Spain thus saw it as its task to preserve these bulwarks of Christianity in a heathen world at all costs. This war would prove that Spain was still a powerful nation. The Moroccans suffered from the handicap that their forces were weaker and untrained; and to make things worse, Sultan Abd-al-Rahman died quite unexpectedly at this moment. After a short period of disorder, he was succeeded by his son, Sultan Mohammed ibn Abd-al-Rahman (1830-1873). Spain officially declared war on Morocco on 22 October 1859. Sultan Mohammed sent his troops to the north of the country, where the Spanish army lay in wait for them. Spain

137 *The Times* 12 Sept. 1859.

138 Minister under Sultan Abd-al-Rahman.

139 *The Times* 12 Sept. 1859.

140 *The Times* 26 Sept. 1859.

gained a decisive victory in the battles of Tetouan and Wad-Ras, and the Sultan capitulated on 23 March 1860. A peace treaty was signed on 26 April 1860, in which the Sultan acknowledged that the cities of Ceuta and Melilla were Spanish and agreed to pay reparations.

Kossuth

In his letter to Ferrier, Schliemann mentioned "the famous speech delivered by the unfortunate Kosuth in the Capitol at Washington in July 1851". Apart from the comment referring to the palm tree flourishing in spite of oppression, this was a remarkable statement. It is true that Schliemann was in the United States in July 1851, but in California and not in Washington. It follows that he cannot have heard this speech personally, and must have been citing from another source.

Schliemann left an s out of the name he mentioned. The person referred to is Lajos Kossuth (1806-1894), a Hungarian nationalist who campaigned for the independence of his country. He and a number of associates proclaimed the republic of Hungary in 1848. He initially became the minister of Finance, and somewhat later the Prime Minister. However, Austria intervened with help from Russia to bring the rebellious part of the double monarchy under the control of Vienna again. Kossuth and his associates fled to Turkey, where they were immediately imprisoned. This brings us to the development mentioned by Schliemann at the beginning of this section. Kossuth was released on 1 September 1851 under pressure from both the English and the American government, and brought to New York on an American naval ship.[141] He arrived on 9 December. The *New York Herald* reported his arrival at length.

> *Eleven o'clock was the hour appointed for the reception of Kossuth at Castle Garden, but it was not until twelve that the landing was made. In the meantime several thousand people were assembled within the spacious hall. Among them were several ladies, the fair sex evidently taking as much interest, and being as desirous of seeing and hearing the great Magyar as the men. At length Kossuth disembarked and then ensued a scene of excitement which it is impossible to portray. Bang, bang, bang went the guns – tap, tap, tap went the drums, Hurrah, hurray! Three cheers! Bang, bang, bang. Hurrah! Three more. There he is! Where? There's Kossuth! Hurrah!* [142]

Lots of applause and a speech by the Mayor. More cheering, so that Kossuth's words of thanks could hardly be heard above the noise.

> *I am half sick, gentlemen, tossed and tumbled about by a fortnight's gale upon the Atlantic's restless waves; my giddy brains are still twirling around in a whirlpool, and this gigantic continent seems to tremble beneath my steps. Let me, before I go to work, have some hours of rest, before I go to work, have some hours of rest on this soil of freedom – the soil of freedom, your happy home.*[143]

---

141 *De Nieuwe Utrechtsche Courant* 10-09-1851.

142 *Weekly North Carolina Standard* December 17, 1851.

143 *Weekly North Carolina Standard* December 17, 1851.

Since our main focus here is on Schliemann rather than Kossuth and his political ideas, I will not go into detail about the rest of his speech. He spoke of the liberation of Hungary and all the oppressed peoples of Europe, but made no mention of palm trees.

If Kossuth did ever talk about these trees, he did not do so in Washington but in his own free Hungarian Republic, which had been in existence from March to September 1848.

Cuba

It was not only in Europe that attempts were made to change political boundaries. Similar expansionist ambitions were to be found in the United States.

Texas declared its independence from Mexico. This was not a peaceful process. The Battle of the Alamo in 1836 made Davy Crockett (1786-1836) and James Bowie (1796-1836) legendary American heroes whose names survive right up to the present day. The independent state of Texas joined the United States in 1845. However, the Texans thought that their state was too small, and wanted to expand its boundaries up to the Rio Grande. They initially tried to buy the land from Mexico, but the Mexicans did not want to sell it. This impasse led to what was known in the United States as The Mexican War, and in Mexico as the Guerra de Intervención Norteamericana. After a series of bloody battles, the war came to an end in 1848 with the Peace of Guadalupe Hidalgo, according to the terms of which Mexico had to cede one-third of its territory to the United States.

The Northern States were not happy with this result, which they saw as an attempt to increase the area of the slave-holding states. The Mexican War may thus be seen as one of the causes of the American Civil War, which broke out in 1861.

This successful expansion led the Southern States to think about annexing Cuba.

Conditions on the Caribbean islands had long been unstable. Revolution against French rule broke out on Haiti in 1791 and lasted for thirteen years until the island declared its independence in 1804. The leader of the revolution, Jean-Jacques Dessalines (1760-1806), named himself Emperor Jacques I. He order the killing of all white men, and passed a law stating that no foreigners were allowed to own land on Haiti. Haiti thus became the first independent black state. Fearing that other peoples would follow this example, slave-holding states broke all links with this new state. The Spanish-speaking eastern part of the island separated from Haiti in 1844 as the Dominican Republic.

Before that, Great Britain had passed the Slavery Abolition Act in 1833, abolishing slavery in all British colonies.

All these events were a cause for concern in the Southern States.

> *The importance to the Southern States to our future relations with the West India Islands cannot be overestimated.*

The condition of Cuba and of St. Domingo is such that they cannot long remain as they are. The Spanish Government cannot much longer hold dominion over Cuba in opposition to the designs of Great Britain, and Haytien affairs are becoming so much confused that the strong arm of some naval power can alone settle them.[144]

Schliemann tried to predict political world trends in his diary, following the example of countless writers of letters to the editor of various newspapers. The example of one such letter given below shows clearly the kind of rumours that were spread in the Southern States about European plots and the predictions about the future made in this connection.

Havana, January 3, 1846.

Dear Sir: I have heard numerous rumors that are floating about in this city, and will give you a few of them. There is no doubt some grand scheme is going forward among the European courts. One of the rumors is, that one of the sons of the King of France is to marry the Infanta of Spain, and take possession of Mexico and give it back to Spain. The court of Madrid will appoint the son of Louis Philippe Regent of Mexico. There appears some truth in this rumor, as we have a large Spanish fleet in the harbor, and a large French fleet is expected here to act in concert with the Spanish. England will have a large number of ships of war to cruise in the Gulf of Mexico.

What share of the spoil England is to receive for her share, Madam Rumor does not say, but only hints at Upper California, or the eastern half of Cuba, say to the longitude of 78, as this part of the island will be nearer her possessions. The court of Madrid does not like the idea of giving England a foothold in Cuba as from this island Spain derives nearly all her revenue, and she knows full well that if England once raises the cross of St. George in Cuba the whole island is lost to her, and in lieu of this will offer to Victoria, Porto Rico.[145]

The Southern States demanded that President Franklin Pierce (1804-1869) should take steps to incorporate Cuba, as a slave-holding state, into the Union. Since the provisions of the Kansas-Nebraska Act stipulated that the inhabitants of the individual states were free to determine whether they should be slave-holding or not, President Pierce had to determine what stance to take in this matter. To throw more light on the issue, the American ambassadors in Spain, Pierre Soulé (1801-1870), Great Britain, James Buchanan (1791-1868), and France, John Y. Mason (1799-1859), met in Ostend to discuss the question. They drew up the Ostend Manifesto, which contained the proposal that the United States should buy Cuba from Spain, but if Spain was not willing to cooperate then it was suggested that the United States had the right to take control of Cuba by force. The manifest was supposed to be confidential, but Soulé revealed its contents. When the proposals became known, this unleashed a flood of criticism from the Northern States and Europe. The Southern States were indifferent to this criticism,

144 *Edgefield Advertiser*. Nov. 5, 1845.
145 *The Guard* (Holly Springs, Miss.), February 6, 1846.

and one senator warned the government of the risk run by Cuba as a slave-holding state if it were not admitted to the Union.

> [Cuba] *in the hands of emancipated blacks, or of foreign Powers determined to emancipate, would be a total loss to the commerce of the world, and a death blow to the extension of Slavery … Already the apprenticeship system is being applied, and slavers are daily embarking for the African coast, to return laden with slaves, who in a few short years will work out their rights to a free citizenship on the island of Cuba. This is the design of England and France, and of course Spain will not hesitate to connive at a scheme she has long since desired herself to accomplish. If this suspected object of the European Powers ever be attained, and Cuba fall into the hands of the emancipados, no one can tell the immediate consequences to the inhabitants; but the ultimate extermination of the Creole population, or their complete banishment from the island, is a result too palpable to require a prediction. We have but to revert to the St. Domingo insurrection, and the Haytien emancipation, and think now of his woolly-headed Majesty, Faustin I, to learn the fate of emancipated Cuba.*
>
> *Success to the filibusters, say we.*[146]

The term "filibusters" used in the above extract was derived from "freebooters" and meant people who used armed force to interfere in the political affairs of other countries without permission from their own government. It was mainly used to refer to citizens of the United States who became involved in revolutions in Latin America in the middle of the nineteenth century. It was only later that filibustering was used in a political sense to refer to members of a legislative assembly who tried to hold off a vote by making endless speeches.

The exacerbation of the tensions between the Northern and the Southern States, leading ultimately to the American Civil War, put an end to the ideas of the annexation of Cuba.

Men and women

Schliemann's diary covering his travels through Spain contained many comments not only on the political climate of the time but also on women, reflecting his own views on the relationship between men and women as well as current opinion on this topic.

As he travelled through Spain, he was repeatedly assailed by memories of a young Canadian woman, the daughter of James Ferrier, whom he had met on the boat from Messina to Alexandria. He mourned her as the great love of his life who had not reciprocated his feelings.

He had little to say about St. Petersburg apart from the fact that it was cold: he had had to wear a fur coat while at his house in the country. What he writes gives the impression that he was a 37-year-old bachelor. Nothing could be further from the truth. Seven years before, in 1852, he had married Ekatherina Petrowna

---

146 *The National Era*. March 23, 1854.

Lyshina (1826-1896). It is true that she only married him for his money, but they had had two children. His son Sergey (1855-1939) was born in 1855 and his daughter Natalya (1859-1869) in 1859, while he was in the Middle East.

> *Zuerst vor allen Dingen muß ich Ihnen zur Geburt Ihrer Tochter gratuliren, – möge sie aufwachsen zu Ihrer Freude und Ihr häusliches Glück vermehren, sorgen Sie nun aber auch daß Sie baldmöglichst das Geschenk von Angesicht zu Angesicht schauen und der langen Strohwittwerzeit Ihrer Frau ein Ende machen, denn es kann Ihnen doch Alles so recht nicht schmecken, wenn Sie an die Entbehrungen Ihrer Ehehälfte denken.*
> *(First of all I must congratulate you on the birth of your daughter – may she grow up to be a joy to you and increase the happiness of the whole family; but now you must take steps to see this treasure face to face and put an end to your wife's long grass widowhood, for you cannot really enjoy life when you think of the privations of your other half.)*[147]

F. W. Hepner (?-1869), who wrote the above to Schliemann on 16 April 1859, had however no idea how bad Schliemann's marriage really was. Schliemann gave a description of his marriage in a letter to Count Fehleisen, who also lived in St. Petersburg. His wife described him to anyone who would listen to her as a terrible tyrant, a despot and a libertine; but none of that was true, it was all her own fault. After her first pregnancy, she no longer wanted him to touch her. Whenever he approached her, she would scream that he wanted to murder her. Because she would not admit him to her bed, he had to steal his two last children[148] from their mother.[149]

It is unlikely that Schliemann was entirely responsible for the state of his marriage. There are indications that his wife despised him because he did not give her noble family enough money. On the other hand, there are indications that he preferred the company of other women to that of his wife. In any case, the fact that he did not mention his wife and children at all in his diary makes it clear that he had a highly dysfunctional marriage, certainly by nineteenth-century standards.

The mere fact that there is no indication in the diary that he visited his wife and children while in St. Petersburg is clear evidence of this.

Of course, the diary was not intended to be read by others, but it is striking that he does not even try to keep up appearances for himself.

The relationship between man and woman was well defined in the nineteenth century. The family was all-important for people in the middle classes. Such families were quite large, with five or six children on average. Each member of the family knew his or her place, and education was aimed at reinforcing this. The father was the head of the family, and all other members of the family and the servants had to obey him at all times. His children had to call him "Sir". His wife was far from being his equal, and also had to obey his commands without question. It is surprising in this connection that Ekatherina Petrowna Lyshina complained

147 Meyer 1953, 96.
148 His second daughter, Nadezhda (1861-1935), was born in 1861.
149 Traill 1995, 31.

Figure 6: The Lovers 1855 – by William Powell Frith (1819-1909)

that her husband was a tyrant and a despot. That should not have surprised her: it was the role that the head of every family was expected to play.

The wife was in charge of the running of the household. Wherever possible, she did not do the household work herself but told the servants what had to be done. The servants were not only there to perform the household work, but as a sign of the prosperity of the family. The more prosperous a family was, the more servants it had.

Education was mainly aimed at teaching the children their place in society and how they should behave to maintain that position. The mother had the task of teaching her children how to read and write, while in more upper-class families the father would teach his sons Latin and Greek. This may be one of the reasons why Schliemann started learning Greek in 1856 and Latin in 1858.[150] When the

150 Schliemann 1892, 23. According to his biography, everything that Schliemann did was directed towards the aim of excavating the site of ancient Troy. This made this document more of a hagiography than a biography; as a result, the reasons he gives for his actions are not always reliable.

children were slightly older, a governess or tutor was often employed to continue their education. Later, the boys – and also the girls in Catholic families – were often sent off to boarding school. Schliemann visited a girls' school in Toledo. Unfortunately, he only described the building and gave no details of the way of life there and what the girls were taught.

Boys were expected to work hard to prepare for a good position in society, while girls were expected to marry as soon as possible. This would protect them against destitution in later life. The threat of the workhouse hung constantly over the heads of people in the lower classes. The big, polluted cities, where the poor lived packed together in unhealthy slums, bred many infections. Polio, measles and tuberculosis were deadly diseases under those circumstances, and if the father had no work the whole family was doomed to a short life of hunger and misery.

In 1798 the preacher and economist Thomas Robert Malthus (1766-1834) published his *Essay on the Principle of Population*, in which he claimed that the population would always tend to grow faster than the economic means of sustaining it. Unless appropriate measures were taken, it would become impossible to grow enough crops to feed the increased population. The impoverished lower classes would continue to expand and would ultimately threaten the existence of the upper classes. It was thus necessary to restrain population growth, especially among the lower classes, and this could only be done by improving moral standards. It was the duty of the upper classes to give an example here.

Many physicians in this period believed that only men had sexual urges. Women were free from this. Only lower-class women had sexual appetites. If upper-class women showed any signs of this, they were considered to be suffering from unhealthy perversions. Sex for women was only intended for the procreation of the species, and should not be indulged in more than absolutely necessary. Since however nineteenth-century women did also have sexual appetites, this led to a double morality where there was a difference between the standards that people expressed in public and what they did in private. As a result, sex outside marriage was regarded as socially unacceptable, especially for women. If an unmarried woman became pregnant, it was obvious that she must have been indulging in sexual relations: she had given way to her passions and transgressed the unwritten laws of society. The poor girl who had been seduced and abandoned, and came to an unhappy end, was the theme of countless novels and plays. A good example is *Adam Bede* by George Eliot (the pen name of Mary Ann Evans, 1819-1880), which was published in 1859. Although this novel deals with an extramarital affair and its disastrous consequences, the precise nature of the sexual acts that led to the downfall of one of the main characters is only suggested rather than described in any detail. This is typical of the way it was considered proper to deal with such topics in the nineteenth century.

> *Hetty turned her head toward him, whispered, "I thought you wouldn't come," and slowly got courage to lift her eyes to him. That look was too much; he must have had eyes of Egyptian granite not to look too lovingly in return.*

"You little frightened bird! Little tearful rose! Silly pet! You won't cry again, now I'm with you, will you?"

Ah! He doesn't know in the least what he is saying. This is not what he meant to say. His arm is stealing round the waist again, it is tightening its clasp; he is bending his face nearer and nearer to the round cheek, his lips are meeting those pouting child-lips, and, for a long moment, time has vanished. He may be a shepherd in Arcadia for aught he knows, he may be the first youth kissing the first maiden, he may be Eros himself, sipping the lips of Psyche – it is all one.

There was no speaking for minutes after. They walked along with beating hearts till they came within sight of the gate at the end of the wood. They looked at each other, not quite as they had looked before, for in their eyes there was the memory of a kiss.

But already something bitter had begun to mingle itself with the fountain of sweet; already Arthur was uncomfortable. He took his arm from Hetty's waist, and said, "Here we are almost at the end of the Grove. I wonder how late it is," he added, pulling out his watch. "Twenty minutes past eight – but my watch is too fast. However, I'd better not go any farther now. Trot along quickly with your little feet, and get home safely. Good-bye.[151]

Although this passage does not contain a single improper word, it is clear that there is a discrepancy between the outer description and the inner reality. Hetty and Arthur must have shared more than a passionate kiss.

One of the clearest examples of nineteenth-century double morality is the case of Charles Dickens (1812-1870). His novels are full of pictures of the ideal family. His own personal behaviour, however, was far from this ideal. When he was 45 he separated from his wife and embarked on a passionate affair with the 18-year-old actress Ellen Lawless Ternan (1839-1914). This affair had to be kept secret, to avoid scandal.[152] Scandal was the worst thing that could happen to a nineteenth-century family. It could lead to the family being ostracised from society and even ending up in the workhouse.

Schliemann had experienced the results of scandal when he was a boy. While his mother was pregnant with her last child, his father had an affair with the housemaid, and when the mother died shortly after, the father "lived in sin" more or less openly with the maid. This was unacceptable behaviour for a minister of the church. The scandal was too much for his parishioners. Schliemann's father was dismissed from his position, lost his parish with its generous stipend, his house and his land, and could no longer afford to pay for his son's education at gymnasium or later at university. The scandal caused by the father robbed the son of the possibility of an academic or intellectual career.

This double morality led to a double image of womanhood. On the one hand there was the healthy woman without sexual appetites, who dutifully looked after the household for her husband and children. On the other there was the degraded

151 Eliot 1859, 117-118.
152 See Tomalin 1990.

woman, driven by passions, who drove men to ruin. Womankind consisted either of Madonnas or of whores.

This duality is reflected in Schliemann's diary. On 5 September he described a letter he had given in Jerusalem to a young Canadian woman he loved. He told her that he would be the happiest man alive if he could be near her for the rest of his life. After writing at length about this letter, he mentioned in a few lines that he had seen a servant girl across the street from the window of his room, and had been convinced that she was a whore. He had called out to her, telling her to come to his room at 10 o'clock at night, and was astonished when she did not turn up.

The large number of prostitutes to be found in the nineteenth century showed the vast gap between preaching and practice at that time. It has been estimated that there were about 80,000 prostitutes in London in 1860, that is 2.6% of the total population of the city. There were many different kinds of prostitutes, young women who sold their favours on the streets, in the harbours, those who worked in brothels and those who plied their trade in theatres. The poet Dante Gabriel Rossetti wrote sympathetically of one such prostitute in his poem *Jenny*

> *Lazy laughing languid Jenny,*
> *Fond of a kiss and fond of a guinea,*
> *Whose head upon my knee to-night*
> *Rests for a while.*[153]

Many in the nineteenth century regarded prostitution as a necessary evil. Men had their needs, which had to be satisfied or they would become ill. Masturbation was not the solution. It was believed that this was a degrading practice: spilling one's seed in this way would lead to loss of vital energy, and those who practiced it habitually would ultimately waste away and come to a miserable end. Since moreover it was desirable to limit the number of children being born, prostitution was the only solution. Nevertheless, many were against prostitution. Prostitutes were fallen women who needed to be saved, and many well-meaning people tried to do so. English Prime Minister William Gladstone was a famous example. It may be asked however whether such campaigns were really aimed at solving moral problems, or whether there were other underlying motives. While prostitution was not an honourable occupation, many women preferred it to factory work. They often earned more, and their working conditions could be more agreeable. If they were not controlled by a pimp or brothel-keeper, they could be independent of men and more or less regulate their conditions of work themselves. As a result, many men and women regarded them as a threat to the established order. Their whole existence threatened the idea of male superiority, and it was feared that it would have far-reaching social consequences if other women came to share their ideas.

Apart from this hypothetical social threat, prostitution was associated with another, more practical risk, the spread of sexually transmitted diseases – in particular syphilis. This infectious disease normally occurred only in lower-class women and prostitutes. This disease was considered to be hereditary, and by

153 Rossetti, Dante Gabriel 1870 *Jenny.*

Figure 7: Daguerrotype from about 1850 – by Felix Jacques-Antoine Moulin (1802 – 1875)

having sex with middle-class or upper-class men and infecting them, these women represented a threat to the social order. By spreading this disease, they involved the upper classes in the resulting social degradation.

Many nineteenth-century women from the middle and upper classes were affected by another mysterious disease, hysteria. While the reasons for this condition are still far from clear, it is a fact that it led to a wide range of complaints and rendered those who suffered from it unable to do much more than lie in bed and rest.

It is clear, however, that the roots of hysteria lay in the relationship between men and women in this period. Men were supposed to be strong, powerful and intelligent, while women were seen as weak, sensitive and irrational. This led to a remarkable world view. Middle- and upper-class white men were considered to have developed to the pinnacle of perfection, thanks to the above-mentioned properties, far superior to the "modern savages" of the working classes and ruling over them by reason of this superiority. The women from these classes, on the other hand, had undergone exactly the opposite development. They were far weaker than their male counterparts, and could not be compared with working-class women, who could work just as hard as their men. The cook of a middle-class family, coming from the working class and working under terrible conditions, was likely to be in excellent health, while her mistress was often a physical and emotional wreck who found it difficult to get through the day. According to medical views at that time, these problems were due to excessive exertion. Dr. R.R. Coleman from Birmingham, Alabama, put it as follows:

Women beware. You are on the brink of destruction: You have hitherto been engaged in crushing your waist; now you are attempting to cultivate your mind: You have been merely dancing all night in the foul air of the ball-room; now you are beginning to spend your mornings in study. You have been incessantly stimulating your emotions with concerts and operas, with French plays, and French novels; now you are exerting your understanding to learn Greek, and solve propositions in Euclid. Beware!! Science pronounces that the woman who studies is lost.[154]

The reading of novels, not just French novels, was considered to be particularly unwise. It could have catastrophic consequences, such as drying out of the womb leading to infertility and even to madness. (The term hysteria is derived from the Greek word for "womb": it was believed at that time that this condition was due to some malfunction of the womb.)

The marquise whom Schliemann saw in the insane asylum in Toledo may have been suffering from hysteria. It is also possible that she was the victim of a typically Spanish complaint. Miguel de Cervantes (1547-1616) stated that his hero Don Quixote lost his reason through the reading of too many romances of chivalry. However, Schliemann's marquise bears little resemblance to a quixotic knight but rather displays a picture of total exhaustion that fits the diagnosis of hysteria much more closely.

Schliemann's suggestion that he should take a 14-year-old girl from her mother in Seville and live with her in Paris is highly offensive to our modern sensitivities. It sounds like paedophilia to us – and her mother, who turns this suggestion down, and her acquaintances seem to have had the same impression. It is doubtful, however, whether this view would have been generally shared in the nineteenth century. Lewis Carroll (Charles Lutwidge Dodgson, 1832-1898), the professor of mathematics at Oxford University, not only wrote the charming story *Alice in Wonderland* for ten-year-old Alice Liddell (1852-1934) but also took photos of her and other very young girls, with the permission of their parents, showing them naked or half-naked. We do not know whether Carroll went further than taking photos, but the fact that the parents allowed him to take nude photos of the little girls suggests that they were not worried by these activities.

The pornographic literature of that period also quite often features young girls.

Finding that two pretty young girls of fourteen and fifteen were living close by,with an invalid mother, whilst their brother was away, being a Midshipman in the Royal Navy, I proposed that Annie should send the Misses Bruce an invitation to spend the afternoon with us, en famille, without the least ceremony, and join us in an alfresco tea party at a little hut in the woods, which formed part of my uncle's estate.[155]

The story continues with descriptions that leave nothing to the imagination. It is clear that Schliemann's revelations in the present diary, though far from pornographic, are equally far from the delicacy shown in the quotation from

154 Cited in Ehrenreich & English 1979, 115.

155 *The Pearl, A Monthly Journal of Facetiae and Voluptuous Reading* No. 7 1880.

George Eliot given above. Schliemann's open-hearted use of language in this fourth part of his diaries may be seen as evidence that he was only writing from himself, and that he did not intend these details to be read aloud to others as had been the case in his description of his first European trip.[156]

The Ferrier Family

The Ferrier family appeared in Schliemann's life during a trip from Messina to Alexandria, which lasted from December 27, 1858, until January 4, 1859.

> *... We were 14 passengers in the first cabin. Among them there is a young lady whom I am very fond of. Her family plans on travelling for 2 years. She comes from Montreal, she is plain, she only speaks English and yet her upbringing is excellent. It seems that ... the aforementioned honourable family's name is James Ferrier.*[157]

He was immediately smitten by the daughter, but nothing came of it, as becomes clear from the diary he kept during his trip through Spain.

That Schliemann felt attracted to the Ferrier family is easily to understand. There were a large number of similarities between the life of the paterfamilias, James Ferrier (1800-1888), and Schliemann, but in contrast to Schliemann, Ferrier had a wife who accompanied him on this trip.

Even if James Ferrier played an important part in the history of Montreal, which brought him an extensive lemma in the *Dictionary of Canadian Biography*,[158] little is known about his family. He was married to a Mary Todd and had at least two sons and two daughters.

To a certain extent there are clear parallels between the lives of Ferrier and Schliemann. Ferrier was born in 1800 in Fife in Scotland where he worked for some years in a commercial establishment. In 1821 he immigrated to Canada where he again worked in a commercial firm. Eighteen months later he opened his own shop. The profit he made thereby he invested rather well in real estate. By 1836 he had accumulated enough wealth to be able to retire and live as a gentleman with a fortune. From this moment on his main activity was looking after his investments. He had large interests in, among other enterprises, banks, insurance companies and railroads. In contrast to Schliemann, he was not a distant investor, but almost always was involved in the running of the firms in which he had invested. So he was, for example, a board member of The Bank of British North America, president of the Montreal Fire Assurance Company, the Montreal Mining Company, the Montreal and Lachine Railroad and the Grand Trunk Railroad – to mention only a few of the companies in which he had invested large sums of money and in which he played a role.

Next to this and perhaps partly in service of this he played, until his death, an important role in Montreal politics. He was, among other things, alderman and mayor. Ferrier was convinced of the importance of good education and was

156 GL Sch A: A1 diary 1846-1847 and Thanos & Arentzen 2012.

157 GL Sch A: A3 diary 1858-1859.

158 Tulschinsky 1982 .

Figure 8: James Ferrier – Library and Archives, Canada

responsible for the fact that he rescued McGill College when it was in serious financial straits, which led to it becoming McGill University.

He was, compared with Schliemann, a religious man. He came as a Presbyterian from Scotland but became a Methodist in Montreal. Here too he played a leading role and contributed generously to the building of a number of churches. He performed all functions in this church that were open to a layman. He even taught Sunday school.[159]

Although Schliemann, in 1859, never wanted to meet the Ferriers again, he did not keep himself to this decision. During his world tour of 1864-1865, on November 2 he visited Montreal. During the day he viewed the city and in the evening he visited the Ferriers.

> *In the evening after dinner I drove to Mr James Ferrier 100 Alexander Street with whom & with whose family I had travelled in 1858 1859 in Egypt & to Jerusalem. He & his wife both of whom look now much younger than they did 7 years ago, highly rejoiced to see their old friend and insisted I should stay with them a couple of days, which however I most peremptorily declined. Then they insisted I should at least come back at 9 in the evening to supper since their daughter Mary now married to Mr James Torrens, would always regret not having seen me. He gave me his & his wife's portraits as well as that of his son, who was with them in Egypt and who is now clergyman. He also made me a present of his biography, from which I see with astonishment what a wonderful man he is; he only carried on a grocer's business for 12 years here and then retired, 35 years old, in 1835, with a fortune of 120/m £ and ever since that time he only works for others being continually elected to places of honor as director of Colleges asylums, as senator for parliament & since 1847 as director of the Great Trunk Railway*

159 Tulschinsky 1982, Bloedow 2001.

and by his great deal of energy and sound reason he gives everywhere the fullest satisfaction and for instance the Great Trunk Railway, which was bankrupt and hardly paid $ 50/m weekly when he became its director in 1859 gives now already $ 150/m weekly. I came back at 9 p.m. when I saw the whole family; his daughter Mary pleased me as much as ever before as also her husband Mr Torrens who has black hair and is a very handsome man; these latter have 2 boys, the one called James Ferrier Torrens and the other James Torrens Torrens. Mrs Ferrier is suffering from deafness. The elder daughter is married to a former clerk of Mr Ferrier of the name of Mr Furlane. They are very nice goodhearted people and I separated from them with sincere regret; but I love them so much that I thought the sooner we separate the better it would be for me. This house is 2 stories high of granite surrounded by a garden with a hothouse annexed.[160]

It is noteworthy that Schliemann here claims that he travelled together with the Ferriers to Jerusalem. But the fact is that this was not the case, and was to a certain extent the main reason for the unhappiness and the split between them that he describes in his Spanish diary.

From the final lines one might conclude that Schliemann still had tender feelings for the daughter and left in great haste since she was already married.

Concerning the biography of Ferrier, which he allegedly gave to Schliemann, I am mystified. I have until now been unable to find even the slightest trace of a biography or autobiography of Ferrier.[161]

Schliemann's precipitous departure was, contrary to what one would expect, not the end of the relation between Schliemann and Ferrier. When Schliemann was in Indianapolis in 1869, to obtain a divorce from his Russian wife, just before he returned to Paris, he wrote a letter to Ferrier from New York to solicit his help in securing portions of the Bible that were translated into Indian languages. Several lines in this letter are somewhat remarkable. As the reason for his stay in the USA he gives the following:

I have only come over here on a short visit to look to my property.

He does not speak about his wife or the divorce, and so it seems certain that he had never told Ferrier that he was married.[162] This view is supported by what he wrote concerning his future plans:

About a week after my return to Paris I intend going to Athens, when I shall probably marry a young Greek lady, who is just such an enthusiast for ancient Greek literature, archaeology, ancient history and geography as I am and there I hope we shall be happy together.

160 Gennadius Library A8, 68-69. See also Bloedow 2001.

161 I thank David Greene, Librarian of McGill University, for the quest that he undertook on my behalf in 2015, one that, alas, did not yield any information.

162 Bloedow 2001 note 64 "... Schliemann's omission in his letter to Ferrier of any reference to the principal reason for being in the USA in 1869. This was no doubt because he was aware of the Christian convictions of the Ferriers, who would certainly not have approved of divorce."

I cannot stand the solitude any longer and least of all in Paris, though I have there always the house full of the most charming society.[163]

If Ferrier ever answered this letter is unknown.

Schliemann's first publication

Schliemann mentioned in his letter to James Ferrier that he had published a brief account of his visit to Petra in *The Times*. Apart from a recommendation for his English teacher, which appeared in the *Algemeen Handelsblad* on 2 November 1842, this is Schliemann's first publication.

During Schliemann's stay in Jerusalem, the British consul James Finn (1806-1872) helped him to visit the Church of the Holy Sepulchre. Finn was consul in Jerusalem from 1846 to 1863. On his return to England, he wrote a book about his travels in the Holy Land during his stay in Jerusalem. I give below some details taken from this book of a trip he made to Petra in 1851, in the interests of the European and American travellers who visited this site. Even at this time, Petra was a considerable tourist attraction. Travellers paid a lot to go there, and the local Bedouins were keen to get their hands on as much of this money as possible. In Finn's opinion, the Bedouins' demands had become excessive.

During the last twenty years there have been many English and other visitors to Petra: but they usually take it in the way from Egypt towards Jerusalem … on hearing that several travellers had been unable to reach Petra even after 'Akabah, on account of hostilities arising between the Alaween and the Tiyâhah Arabs, or on account of the exorbitant demands of money made by the former of these, I thought the time had arrived for me to show the practicability of getting at the wonders of Petra from Jerusalem, under escort of the Jehälcen Arabs near Hebron.[164]

Finn travelled to Petra in April 1851, and thanks to the agreements he reached with the local Bedouins Schliemann was able to make the same trip in 1859. For Christians, travel through this region was like travelling through the Bible.

At Beni Naim is the reputed sepulchre of the Prophet Lot, according to the Moslems; that of his daughters being on an opposite hill at no great distance. This village commands a grand prospect of the Dead Sea, although there is no view of the kind from all the country around. Is not this the place whence Abraham, after the departure of the angels, saw the smoke of Sodom and Gomorrah rising as the smoke of a furnace?[165]

It was spring, and the desert was at its greenest.

It was a smooth, pebbly plain, dotted with shrubs, having lines of chalky hills to the south-west, for which our people had no other name than Jebel el Ghurb, or the "western mountain." The whole scene was that of a mere desert; no creatures

163 Gennadius Librarry BBB 28 164, Bloedow 2001, 18-19.
164 Finn 1877, 289.
165 Finn 1877, 291.

were to be seen or heard but ourselves. No Turkish authorities ever intrude into this purely Arab wilderness ... Not far from this must have transpired the incidents recorded of Hagar and Ishmael, – incidents that might have occurred yesterday, or last week; for a few thousand years count but little in so primitive a region.[166]

However, the local population was not so primitive that they did not understand the value of tourism – though the methods they used to extract money from travellers were rather crude by modern standards.

Ascended a series of precipices, and, at some elevation, met two young English gentlemen, with a pair of double-barrelled pistols shared between them, and their fingers ready on the triggers. They had a tale to relate of grievous exactions made by the Fellahheen of Petra, – which, however, seemed to me, by their account, to have been brought on unconsciously by themselves, in having taken an escort of Tiyâhah Arabs from Nukh'l instead of the Alaween; and they informed me that a clergyman from Cambridge was still detained there, as he refused to comply with the excessive demands of the people.[167]

The clergyman's refusal was pointless. Finn met this gentleman the next day. The Arabs had already let him go, after extracting the last cent from him, and he was now on his way back to the civilised world. Despite problems of this kind, a visit to Petra was an experience that was not to be missed.

Entering a valley of red rocks, much streaked with blue in wavy lines, the first work of antiquity that met our view was a square turret on each side of the road. Then we passed some tombs, or chambers, cut into the massive red cliffs with architectural cornices, pediments, and pilasters, some of them very handsome. Next was what Laborde[168] *marks in his map as "the solitary column." It is standing solitary; but then near its base lie other columns of the same edifice, with the circular slices (or drums, as architects term them) that composed them, scarcely disturbed as they slid down in falling. In five minutes more we halted for the night close to what Laborde designates the Acropolis, where a pile of fine building lies prostrate, and the columns on the ground, in their segments, still touching each other. At the foot of this heap stands what is named the Palace of Pharaoh; and our station within it appeared, from the black relics of fires there, to be a frequent resting-place for travellers.*[169]

This may not be the most sublime description of Petra that has ever been published, but Finn was not there to write a travelogue. He had come to negotiate with the local Arabs about this new route to this famous site.

---

166 Finn 1877, 197-298.

167 Finn 1877, 305-306.

168 Petra was rediscovered in 1812 by the Swiss traveller Johan Ludwig Burckhardt (1784-1817). In 1828 Louis Maurice Adolphe Linant de Bellefonds (1799-1883) and Léon de Laborde (1807-1869) visited it. They stayed long enough to make pictures of most of the buildings. In 1830 Laborde published his impressions as *Voyage de l'Arabie Petrée.*

169 Finn 1877. 311-312.

... the indigenous peasantry of Wadi Moosa were gathering around our tents from different directions. They had not been prepared for the reception of guests arriving from the north, i.e., Jerusalem, as travellers usually came from Akabah or Sinai, through Nukh'l.

Our Arabs, both Jehâleen and some strangers, set to making themselves comfortable. There arrived a large body of the Fellahheen, headed by Shaikh Sulimân es Saïd, a ragged and ugly crew, he as dirty as the rest, but strutting about in a robe of bright scarlet.

Then commenced the negotiations and disputes between them and ours; noise and menace speedily ensued, alternated with diplomatic manoeuvres, for our champion, Selâmeh, was an able practitioner in such matters, at least he had an able reputation for it. The stormy scenes were not concluded till late in the night, and they ended by an arrangement to the territorial owners as had been hitherto claimed from those arriving under Alaween escort from Nukh'l or 'Akabah; and this agreement I ratified orally, as writing or sealing would have been altogether out of place there ...

April 7th. – Sunrise ... An inundation of strange Arabs from the desert had arrived during the night, and it was computed that there were not less than two hundred guns round our tents, while our party had not more than five, with a few pistols. We were hemmed in by the newcomers, and the crags over us were occupied by men with guns laid in position between crevices. Some men were scattered about, shooting at birds; but it seemed to me their real objet was rather the making of signals.

These people were 'Ali Rasheedd's branch of the Alaween, from a district not so distant as 'Akabah. Our Jehâleen party looked very insignificant among them; they had evidently not expected this turn of events.

As soon as we Europeans showed ourselves after breakfast, the Fellahheen rushed forward to serve as guides in exhibiting the curiosities ... On their return, the Fellahheen were rapacious in demands for remuneration of their services, but were at length content. This was the signal for the others to take their advantage. They wanted toll to be paid for crossing part of the desert on which they thought the Jehâleen had no right or precedent for bringing strangers. So, our preparing to leave the ground, they rushed up the bank, secured commanding points for their guns, and thus exacted their fee. The screams and hubbub were at length terminated by some small backsheesh, (to our surprise, how little was required,) and we all marched away in northern direction, the opposite to that of our arrival.[170]

That is how Finn opened the route from Jerusalem to Petra for travellers from the West. When he wrote his book in 1877, he added that since he had spoken with the local Arabs many travellers had taken this route without any problems. By 1859, however, this route seems to have become less trouble-free than Finn expected. Schliemann and his fellow-traveller W. Marshall felt obliged on their return to Jerusalem to warn other travellers of the problems that might await them

170 Finn 1877, 313-316.

during a trip to Petra. They did this by means of a letter to the editor of the *London Times* that was published on 27 May 1859.

A VISIT TO PETRA

TO THE EDITOR OF THE TIMES

Sir,- As there have been considerable difficulties of late in the way of visiting Petra, and as we have reason to believe that we are the only European travellers who have been there for nearly two years, we think it may be of service to others if we state in your columns the result of our experience.

We, with a Russian gentleman and a young English boy, started from Jerusalem on the 26th of April, accompanied by an experienced dragoman, and under the guidance of the Jehâlin Bedouins. Their fine old sheik, the Hadjee Abao Dahuk, escorted us himself, and from him and his men we received the greatest civility and attention throughout.

We met with no difficulties whatever in reaching Petra, and entered it on the evening of the 1st of May, having ascended Mount Hor by the way. We remained in Petra the whole of the next day, and the Wady Mousa Arabs and their sheik seemed quite glad to see us, and acted as our guides throughout the day, taking us whither we would, and giving us no trouble beyond the usual importunity for baksheesh, an importunity to which everyone has got well accustomed after being a little time in the East.

The following morning the Arabs gathered in considerable numbers, and there was a good deal of noise and squabbling among themselves. One of them, getting excited, suddenly seized the dragoman, demanded that he should open his canteen, and gave him a few cuffs on the head when he refused; but some of the others immediately interfered on the dragoman's behalf, and drove off the assailant. Presently the Sheik of the Benjazi tribe came up with some of his followers, and quietly made the Wady Mousa Sheik disgorge the whole of the baksheesh (about 11l.) which he had received from our dragoman, and, as a disturbance among themselves was evidently brewing, we deemed it more prudent to take our departure, which we did perfectly unmolested. Indeed, one or two of the Wady Mousa men volunteered to guide us down to the valley, and did so.

We believe that with an experienced and discreet dragoman, and with some forbearance and judgment on the part of travellers themselves, there need be no obstacles in the way of enjoying the glorious scenery and unique grandeur and beauties of Petra for one whole day, at all events, as we did. It may be well to state, however, that the Sheik of Wady Mousa informed us that travellers coming from Mount Sinai must go round by Nakel, as he could have no dealings with Sheik Hussein, who used formerly to conduct travellers from the South.

It would be worthy of consideration, also, whether, as the Benjazi tribe are evidently more powerful than that of Wady Mousa, it would not be advantageous to place one's self under their protection on entering Petra. Their sheik told our dragoman that if sent for by European travellers he would come down to the plain with a body of his men, guard them during their stay in Wady Mousa, and conduct them back in safety again to the plain. Of course, this would entail a handsome baksheesh, but we think it would probably be the wisest course.

We remain, Sir, your obedient servants,

Jerusalem, May 10. W.M.
G.H.S.[171]

The Satrustegui family

Schliemann wrote on 31 July 1851 in the travel diary describing his stay in California that he had engaged two employees, an American and a Spaniard. They would be able to deal with the two main languages spoken in Sacramento without any trouble. The Spaniard was "Miguel de [sic] Satrustegui native of San Sebastian in Biscaya (Old Spain)".[172] When Schliemann left California in April 1852, he transferred control of his bank to these two employees.

The Banking business hitherto carried on by Henry Schliemann & Co., will in future be conducted by Messrs. Grim & Satrustegui ...[173]

This partnership was short-lived, however. Grim wrote to Schliemann on 13 October 1852 to inform him that the partnership had been dissolved.

Mr. Satrustegui and myself have departed some month since. He is now in Sonnora in the office where his brother were to be and his bro. has gone to South California on sure speculation.[174]

I have not been able to discover what happened to Miguel Satrustegui after that. However, his Spanish origins appear to have influenced Schliemann. Schliemann believed that his arrival in America had made him an American citizen, and that as a result he would be unable to return to Russia. He did not want to remain in California, or apparently to return to Germany, so he was looking for somewhere else to live. He thought for a while of going to live in Cadiz. However, a Russian friend assured him that he would have no difficulty returning to Russia and so he ultimately decided to go back to St. Petersburg.[175]

171 *London Times* May 27, 1859. W.M is W. Marshall and G.H.S. is Gospodin (Mr.) Heinrich Schliemann.
172 Thanos & Arentzen 2014, 166-167.
173 Thanos & Arentzen 2014, 64.
174 Thanos & Arentzen 2014, 65.
175 Thanos & Arentzen 2014, 79.

While I have not been able to discover any correspondence between Schliemann and Satrustegui, he did not forget his former employee. He visited Satrustegui's relatives during his travels through Spain. I have not been able to find any information about Miguel, but some facts are known about his brother, Joaquín Marcos Satrustegui (1817-1885), whom he replaced in Sonnora.

France invaded Spain in 1823, with support from Russia, Austria and Prussia, in order to prevent the adoption of a liberal constitution. This development allowed Ferdinand VII to regain power. The Satrustegui family were exiled to England and lived there for ten years. During this time, the two boys learned fluent English. The family returned to Spain in 1833, on the death of Ferdinand VII. Joaquín began his military career in that year, at the age of 16. He worked as a translator for the British delegation in 1835, during the first Carlist war (1833-1839). When the California Gold Rush began in 1849 , he and his brother went out there to seek their fortune. He and a partner set up their own business in San Francisco in 1850. It is not known precisely what they did, but they seem to have built up a good reputation, as he was appointed Spanish consul there the same year. A year later, in 1851, he also became consul of Parma there. He joined the diplomatic service in 1864, serving as Spanish consul in Newcastle upon Tyne, Algiers, Montreal, New York and London. He was raised to the nobility in 1876, with the title of Baron de Satrustegui.[176]

Unlike Schliemann the Satrusteguis, like many other Spaniards, belonged to the class of emigrants who retained links with their fatherland throughout their lives.

The ascent of Vesuvius

It has been a popular pursuit among critics of Schliemann since 1990 to look for discrepancies between the various versions of the stories he tells. The idea behind this approach is as follows: if it can be shown that Schliemann was not telling the truth about certain details of his personal life, such as the way he became an American citizen, then maybe he is an unreliable witness and cannot be trusted when he tells us about his archaeological discoveries such as the great treasure he found in Troy.[177] I will not go into these and similar suggestions in detail here, but I cannot avoid them completely because this diary contains the second version of a story he had recorded previously. This concerns his ascent of Vesuvius on 10 December 1858, which is described in Italian on pages 41 – 44 of the third diary. Less than a year later, on 1 September 1859, Schliemann recorded in Spanish the version of this event that he had told his Spanish teacher. These two accounts differ substantially in tone: the first is a factual report, while the second is a story with embellishments. This can be seen for example from the way he describes the miserable horse he hired for the ascent. By calling this horse "a true Rosinante", he makes it immediately clear to Spanish readers – and indeed to readers from many other countries throughout the world – who are familiar with the story of Don Quixote what kind of animal it is.

176 For further information see Agirreazkuenaga 2012, 71-89.

177 See *e.g.* Traill 1993.

Figure 9: Vesuvius, eruption cone and crater. Das Buch der Geologie oder die Wunder der Erdrinde und der Urwelt. 1855 Leipzig

Apart from a large number of minor details, the two versions of this account differ in two major respects. When Schliemann mentions two men who assisted him during the ascent, in the first version he states that they give him a helping hand while in the second he writes that they had a sort of sedan chair with them. In the Spanish version, Schliemann makes the ascent more of an adventure, and his own role in it more heroic, than in the Italian version. He reinforces the image by adding that he rejected the offer of being carried in the sedan chair but climbed the mountain unassisted – which evidently not everyone could claim.

The second big difference between the two stories is that in the Italian version he said that he made a souvenir of the ascent from some molten lava and some coins, while in the Spanish version the souvenir was made by others. In this case, Schliemann makes himself out to be braver in the first version. Since the description of the making of this souvenir sounds more credible in the Spanish version, it may be asked which of the two versions is closer to the truth. It is impossible to tell, since he wrote both versions for himself and not for a wider audience. In my opinion, the fact that he wrote two different versions of the same story indicates not that he liked tampering with the truth but that he enjoyed telling stories.

In order to allow readers to compare the two versions of the story, I give the translation of the Italian account of the ascent of Vesuvius below.

This morning I took the 7 o'clock train to Resina. On arrival, I hired a horse and a guide for the ascent of Vesuvius. The route initially follows a wide road paved with large stones, but later it winds through the vineyards up to the lava, which destroyed many fine vineyards last year.

The road resembles that at the Isthmus of Panama, since many of the paving stones have been destroyed by rain storms. As a result, it is difficult to ride over it on horseback, especially with a horse like mine that is hardly able to bear my weight.

After that, we crossed enormous fields of lava, which varied in colour according to the time of the eruption: the lava from the old eruptions is grey, that from more recent eruptions is brown and the most recent black. The rivers of solidified lava assume the most fantastic forms. Climbing and descending over the mountains of lava, we finally arrived at the Hermit's Mountain or the royal observatory. From here to the foot of the volcanic cone, the road was better. We dismounted and began to climb the cone, first of all over the ash in which our legs sank up to the knees. We were joined at this point by two men who went up in front of me and stretched out their hands towards mine, offering to attach me to a hook with which they could pull me up. But knowing that they would charge me an exorbitant price for their assistance, I declined to make use of their services and continued to climb on my own. After struggling through the ash for half an hour, we started to climb over the lava that covered the cone in small fragments. From this point, the going became easier but more dangerous: if I had fallen, I would have broken my neck. We finally reached the summit, where I felt amply rewarded by the spectacular view of the crater and the vista over the fields and towns below us. The surface of the summit appeared at first sight to be covered with verdure, but on closer examination I saw that it was all yellow and green sulphur under which the water boiled and bubbled in thousands of little pools. I approached the crater, but could not see either where it began or where it ended because of the clouds of smoke it emitted. I threw many large fragments of lava into it, but it is so deep that I could not hear them hit the bottom. The guide had told me that this was the main crater. I wanted to walk around it to convince myself of the truth of his statement. All went well for a certain distance, but then we became enveloped in a cloud of smoke that became thicker with every step we took. We could no longer see one another; in fact, we could not even see our hands in front of our faces. The smoke was composed of suffocating sulphur vapour and we could hardly breathe in it. We were very afraid that we were too close to the crater and would fall into it, or that we would fall into the pools of boiling water we could hear boiling away under our feet. We were only separated from them by a thin crust of sulphur and lava. Deep cracks appeared in the ground with each step we took. We threw ourselves on the ground and crept forwards on our hands and knees, keeping in contact by shouting since we could not see one another. We held handkerchiefs in front of our nose to reduce the suffocating effect of the sulphur fumes. We expected every minute to be our last and had given up all hope of escaping from this predicament. We crawled backwards and forwards for at least two hours until by a lucky chance we found the way back and arrived at a spot from which we could see some distance ahead. We quickly stood up and reached the spot where we had left the villainous servant[178] *with the eggs and wine. I agreed with him a price of 5 carlini for a bottle of wine, 4 eggs and a loaf of bread and it did not take him long to boil the eggs in one of the thousands of pools of boiling water to be found at the summit. The water must have been brine, for the boiled eggs were perfectly salty and we did not need to add any extra salt. My joy at being safe and the strong wine I had drunk made me descend the cone too fast and though I had given the guide an arm for safety's sake I got separated from him through the speed with which I was walking. I could no longer control myself. I ran downhill a couple of paces, fell and tumbled head over heels down through the mixture of ash and lava. It might be thought that I would have broken my neck, and my arms and legs, in the fall, but as luck would have it I escaped with a bruised left hand and a painful thumb. I cleaned up my clothes as well as I could, and when I had arrived safely at the foot of the cone I mounted my horse and returned. When I got to the spot on the road near rivulets of molten lava I dismounted, tied up the horse and used a long stick to take a sample of the red-hot lava. With the help of two servants, I put a couple of small coins in the lava. I wanted to keep them as a souvenir of Vesuvius, but when I took them out of my pocket one of them fell on the paved road and broke. I was very pleased with the servants, who did their best to anticipate all my wishes. They did all in their power to assist me, and were satisfied with a very small recompense. In Resina I hired a carriage for 2 carlini to take me to the Post Office in Naples.*

While it is impossible to determine which of the two accounts is closer to the historical reality of Schliemann's adventure, it may be asked how much of the story he made up himself. Was it really as dangerous as he suggested to get so close to the crater? The English geologist James Logan Lobley (1833-1913) visited Vesuvius in March 1868. He published a brief description of his visit later the same year, and a fuller account entitled *Mount Vesuvius: A Descriptive, Historical and Geological Account of the Volcano and its Surroundings* in 1889. He had the following to say about the risk of climbing Vesuvius:

On the question of danger it may be admitted ... that a considerable amount of danger to observers undoubtedly exists ... yet the fatalities that occur are almost always those which befall tourists or others, most probably on a volcano, while active, for the first time in their lives ...[179]

In general, the guides who accompanied tourists during their ascent were well aware of the risks.

At length [after a long walk] *a rounded surface indicated the summit of the long slope, and in a few yards more an almost flat plateau, or rather terrace, was reached ... The terrace was perhaps 20 yards across, with a slope very slightly inclined towards the exterior edge, and from it arose the new cone with the actual eruptive craters ... Higher than the terrace no guide would go ...*[180]

178 Schliemann writes lazzaroni. This is read as lazzarone: scoundrel. Since Schliemann indicates the third time that he uses this word that he is highly satisfied with his lazzarone, we have translated this term as 'servant'.

179 Lang Neil [ca1902], 47.

180 Lang Neil [ca1902], 48.

Logan Lobley went further on his own and survived the trip without major problems, but not everyone was so fortunate.

> *The outbreak of 1872 was attended with a circumstance of tragic interest. A party of excursionists, under the conduct of inexperienced guides, were caught by a torrent of lava. About a dozen were killed, the bodies in some instances not being recovered.*[181]

The future archaeologist

When Schliemann in 1881 wrote his main work on his excavations at Hissarlik, which he had identified as the site of the ancient city of Troy, he stated that he was beginning the book with an autobiography.

> *If I begin this book with my autobiography …* [this is] *from a desire to show how the work of my later life has been the natural consequence of the impressions I received in my earliest childhood; and that, so to say, the pickaxe and spade for the excavation of Troy and the royal tombs of Mycenae were both forged and sharpened in the little German village in which I passed eight years of my earliest childhood.*[182]

In other words, he wanted to convince his readers that his whole life had been marked by his archaeological ambitions. It is interesting to see whether there are any signs of this in the diary he kept about his travels through Spain.

He described in detail how he had told his Spanish teacher in Madrid about his ascent of Vesuvius. At about the same time, he had also taken the train from Naples to Pompeii, but he did not mention his trip to this legendary Roman city; apparently, his adventures on Vesuvius had made the most impression on him when he was there in 1858.

The series of eruptions whose effect Schliemann had experienced had started on 30 April 1855. A new eruption on 1 May was clearly visible from Naples.

> *It was a lovely night, without a cloud, and the moon as bright as in the month of August … there was the mountain standing out of the plain against the blue starry heavens, and a broad stream of lava slowly rolling down its cone, and from the summit a thick cloud of smoke spread over it in the shape of a gigantic pine tree, the lower part of which, from the reflected fire, was red, while higher up a thick black layer of smoke, with a white top, lay curling over like the foliage of the tree. The lava was streaming in one broad flood from about half way up the mountain's cone, whence, when it got to the bottom, it seemed, from the ground being less steep, to crawl along until it reached the edge of the old crater that forms the vacuum between Vesuvius and Monte Somma. Here it found itself arrested on the brink of an abyss some 200 or 300 feet wide; and, from the distance we were at, we could see the stream of fire rushing over like a vast cataract, with immense boulders of red hot rocks hurled over, and chasing each other down until the ground got more even, and the speed slackened. From this point the lava crept on, but so slowly, that it was*

181 Lang Neil [ca1902], 48.
182 Schliemann 1881, 1.

difficult to mark its progress; and the effect was beautiful as it burned the trees it met in its course, which flamed up with a bright blue light like Roman candles ... [we went] *to see the eruption nearer ... we proceeded, both of us dressed as we were, in thin evening shoes and light apparel, until we reached the summit of the mountain, after a severe tug of two hours. However, we were well repaid, and I shall never forget the scene, as we stood within a few yards of the cataract of fire which I mentioned before. A strong wind blew the smoke over Monte Somma, but the heat was terrible, for we were pretty near it, of which you may judge when I tell you that I lit a cigar from the edge of the torrent as it rolled over the edge of the mountain.*[183]

The eruptions followed one another in quick succession, leading to the sudden creation of two new craters on 18 May. It was a great spectacle that was visited by hundreds of people in the course of time. Every self-respecting traveller in the neighbourhood felt obliged to climb the active volcano. As is apparent from Schliemann's account, there were so many tourists that it became worthwhile to make and sell souvenirs on the spot. This early tourist industry was stimulated by the visit of King Ferdinand II of the Two Sicilies and his wife.

Vesuvius was more or less quiescent at the time of Schliemann's ascent, though it was still spectacular enough to attract tourists and amaze them. An ascent was an adventure, and as dangerous as each traveller wished to make it. The volcano did not remain quiescent for long: a new eruption broke out in 1860.

If Schliemann had become a geologist, this episode could have been seen as a precursor of his subsequent work, but he became famous as an archaeologist, not as a geologist. It is noteworthy that Schliemann's account of his ascent of Vesuvius makes no mention at all of the famous eruption of 79 AD that led to the destruction of Pompeii, even though the ruins of this ancient city had attracted countless tourists and antiquarians since 1763. Schliemann stated in 1881 that the excavations at Pompeii had been one of the inspirations for his work at Hissarlik.

Through my father was neither a scholar nor an archaeologist, he had a passion for ancient history. He often told me with warm enthusiasm of the tragic fate of Herculaneum and Pompeii, and seemed to consider him the luckiest of men who had the means and the time to visit the excavations which were going on there.[184]

His remarks about the sexual mores in Pompeii as reflected in the murals there, and his comment that the houses in Cordoba reminded him of those in Pompeii, make it clear that he had visited Pompeii. In both cases, however, these remarks are made in passing and do not reveal any special interest in Pompeii as such.

Schliemann learned some Latin and Greek in his youth from a tutor, in preparation for his studies at the gymnasium (which in fact he only attended very shortly, because of the family poverty I have already mentioned, *e.g.* on page 9). It is not clear how much progress he made in his study of these languages; in any case, it was not enough to enable him to communicate on an equal footing

183 *The Times* 18 May 1855.

184 Schliemann 1881, 3.

with the classically educated elite of his time. In order to bring about a change in this situation, he took lessons in ancient and modern Greek in St. Petersburg in 1856 after the end of the Crimean War, when the market collapsed. He devoted two years to these studies, and started to learn Latin again in 1858.[185] The Greek passages in the diary he kept during his travels in the Orient shows that he had a good command of this language. It is not so clear how much progress he made with his Latin.

The diary describing his travels through Spain includes a passage from the *Iliad* that he used, as many intellectuals from that time would have done, as a comment on contemporary affairs. He wanted to say that, just as Troy had fallen, England would one day fall. The Greek quotation he used for this purpose contained a number of errors:

> *ἔσσεται ἦμαρ ὅτ*[ε] *ἄν ποτ᾽* [ω]*λώλῃ Ἴλιος*
> *ἱρὴ* [n]*αὶ λαὸς ἐϋμμελ*[o]*ί*[o] *Πριάμοιο.*

He translated this as: The day shall come that holy Ilios shall fall, just as the people of the well-dressed Priam.

The passage he is referring to here is Iliad VI, 448-449[186] where Hector says to Andromache:

> *ἔσσεται ἦμαρ ὅτ᾽ ἄν ποτ᾽ ὀλώλῃ Ἴλιος ἱρὴ*
> *καὶ Πρίαμος καὶ λαὸς ἐϋμμελίω Πριάμοιο.*

The day shall come that holy Ilios shall fall, together with Priam and his lance-bearing people.

It is clear that Schliemann must have read the *Iliad*, but he cannot have known long passages of it by heart, word by word. I doubt very much whether this can be taken as an indication that he was particularly interested in Troy at this time, and even less that he wanted to excavate it. It seems to me that he is simply trying to follow the custom common among intellectuals at the time, of making a point with the aid of a classical quotation since he felt that this would add weight to his predictions.

He used the word "archaeologist" once in this diary, but in a totally different sense from that commonly employed today.

> *The only thing that could interest an archaeologist here is the archives at Aragon.*

He was referring here not to excavated objects but to document, using the word "archaeologist" in the old-fashioned sense of "antiquarian", a student of antiquity. He did not seem to be aware that in 1859 the term "archaeologist" was usually used to refer to someone who studied the material remains of classical antiquity.

---

185 Schliemann 1892, 21-23.

186 I would like to thank Reinhard Witte for pointing out this passage to me.

Wout Arentzen

Schliemann's dates and whereabouts

Schliemann gave some of the dates mentioned in his 1859 diary according to the Gregorian calendar, and some according to the Julian calendar that was still used in Russia at that time. The Julian calendar was 12 days behind the Gregorian calendar. It is clear from the dates he used that he had a marked preference for the Gregorian calendar at this stage in his life.

The dates in **bold** type are those given by Schliemann; those in italics are added for the sake of clarity.

August 1859

Saturday **27**	**Bordeaux**
Sunday 28	
Monday 29	
Tuesday 30	
Wednesday 19/31	**Madrid**

September 1859

Thursday 20/01	
Friday 21/02	**Madrid**
Saturday 03	
Sunday 04	
Monday 24/05	**Madrid**
Tuesday 06	
Wednesday **26/07**	**Madrid**[187]
Wednesday 07	**Madrid**[187]
Thursday 08	
Friday **28/09**	**Madrid**

187 Schliemann gives two dates for this day

Saturday **29/10**	**San Ildefonso**
Sunday 30/11	**El Escorial**
Monday 31/12	**El Escorial**
Tuesday 13	
Wednesday 14	
Thursday 15	
Friday 04/16	**Cordoba**
Saturday 05/17	**Seville**
Sunday 06/18	**Seville**
Monday 07/19	**Seville**
Tuesday **08/20**	**Cadiz**
Wednesday 21	**Cadiz**
Thursday **22**	**Cadiz**
Friday 23	**Algeciras**
Saturday 24	**Gibraltar**
Sunday 13/25	
Monday 14/26	**Malaga**
Tuesday 27	
Wednesday 28	
Thursday 29	
Friday 30	

October 1859

Saturday 01	**Granada**
Sunday 02	
Monday 03	
Tuesday 04	**Malaga**
Wednesday 05	
Thursday 06	**on board the "Egyptien"**
Friday 07	**Alicante**
Saturday 08	

Sunday 09

Monday 10 **Barcelona**

Tuesday **11** **on board**

Figure 10: The route of Schliemann's travel

Wout Arentzen

Spanish currency, weights and measures

Spain, like all other European countries, had its own currency until 1850 when it changed to a decimal system. This change was gradual, and when Schliemann visited Spain in 1859 both currencies were in use alongside one another.

The Real de plata fuerte was introduced in 1737. This was divided into $2^1/_2$ Reales de vellón or 85 Maravedis.

A coin worth 8 Reales de plata fuerte was also called a Spanish dollar or Peso, and was legal tender in the United States from 1857.

Sixteen Reales de plata fuerte were 1 Escudo.

The Real de vellón became the standard coin in 1808, but the monetary system remained otherwise unchanged. This system comprised the following coins:

In copper: 1,2,4 and 8 Maravedis.

In silver: $^1/_2$, 2, 4 and 8 Reales; the 8 Reales coin was also known as a Spanish Dollar or Peso.

In gold: $^1/_2$, 2, 4 and 8 Escudos.

In addition, the Catalan Peseta circulated throughout Spain. This had a value of 4 Reales and was divided into 6 Sueldos, 4 Quartos, 8 Ochavos or 12 Dineros. Five Catalan Pesetas were one Duro.

The first step towards a decimal monetary system was taken in 1850. The Real de vellón was divided into ten and was now called simply a Real.

All the above-mentioned coins were in use when Schliemann visited Spain.

Apart from Spanish coins, Schliemann also mentioned Francs, Roubles and Kopeks in his diary.

The system of weights and measures was even more chaotic than the monetary system, and displayed large local differences. The conversion table given below is based on average values, since Schliemann never specified the source of the various units he quoted.

Punto	0.16 mm
Línea	1.94 mm
Pulgada	23.22 mm
Pie	278.6 mm
Vara	83.6 cm
Paso	139 cm
Milla	1393 m
Legua	4.2 km
Almund	4.625 litres
Fanega	55.5 litres
Arroba	14.7 kg
Centena	46 kg
Libra	0.46 kg
Adarme	5.4 gr
Tomine	1.8 gr

Maaike van Asch and Wout Arentzen

The translation

We have done our best to make the translation of the entries in Schliemann's diary accessible to readers by concentrating on the content rather than on the literal meaning of the words. We have avoided the excessive use of synonyms in Schliemann's original text, and where Schliemann makes mistakes we have translated it into what we thought he meant to say. Where the meaning of a particular sentence is unclear, we have kept the translation as close to the original as possible.

[1] Bordeaux 27 August 1859

Yesterday I left Paris at 10 past 9 in the morning, and arrived here at 11 o'clock in the evening. I traveled through beautiful countryside dotted with vineyards. Unfortunately they predicted no good and bountiful harvest for this year; the disease[188] manifests itself everywhere again and the only vineyards saved are the ones that have used sulphur. I passed Orleans, Tours and Poitiers without having a closer look, since the train would not have waited. Before my departure I went to the Office of Rothschild to add the addresses of Madrid and Lisbon to my letters of credit and I took this opportunity to ask if the last Russian loan[189] had gone well. They answered that it had been offered with a discount of 0.25 percent and that it could be bought even cheaper, since there is a lack of buyers. I also inquired about news of the Haber credit guarantee fund, but of that they knew nothing and they advised me to try the pawn shop.[190] There, however, they could not give much information either. I believe that this company was bankrupt before it collected sufficient money. The braggart Haber had falsely acclaimed itself in our city[191] and all the money was all ready to be handed over.

Everywhere in Bordeaux you see evidence of the mixing of Spanish blood with the French and I am charmed by the stylishly dressed ladies with their headdresses wrapped around their heads. The city has beautiful buildings, but it offers nothing interesting apart from the theatre, which is built of large stones and which is surrounded by the pillars of the Botanical Garden, the equestrian statue of Napoleon III erected by the grateful city, the bridge over the Garonne and a few public squares. The exhibition hall, on the other hand, is very beautiful. The building is built from wood. All kinds

188 At the end of the 19th century, phylloxera Daktulo-vitifolia arrived in Europe from America. Around 1870 this insect destroyed about 70% of all French vineyards. The solution was grafting European grape varieties onto the roots of disease-resistance American varieties.

189 Schliemann used the word loan, but he probably meant capital stock or bonds.

190 Schliemann used the term 'Credit of Jewelry'.

191 St Petersburg

of objects are exhibited and some of them show how advanced the art is. There are pocket watches, large bells, all kinds of clothing, tapestries, sewing machines, paintings, hats, portraits, undergarments, shirts and shoes. I was especially charmed by the money vaults that seem capable of withstanding the efforts of any thief, as the lock is equipped with a secret mechanism.

The river offers a pleasant sight, with its many fortifications and fine granite embankments. Those who have come from Paris and who have no business matters to attend to here, will find it excruciatingly boring. I have enjoyed myself in Paris and I hope to enjoy myself much more during the winter when all the scientific circles are open and I can stimulate my mind as much as I wish.

The fortifications are filled with countless cafés; there are many seats out front and at the windows and there one sits and drinks coffee liqueur. [2] There one reads the newspapers and watches those strolling by. The festival given by Napoleon and the entrance into Paris of the troops who had fought in Italy was very spectacular. The fireworks exceeded anything I've seen before. If I were female, I'd be quite happy to always live in Paris, because there one can live as one pleases, without strife and sadness. No one interferes with one another, and whether you wear a bag on your head or an expensive hat, it is all the same.

I believed that I would find giant wine bodegas here, but I was mistaken. In any case, I have not seen anything so far. Thanks to the proximity of the sea, the climate is less hot here than in Paris and I think it's also healthier.

San Sebastián 28 August 1859

I left Bordeaux at 6 this morning and arrived in Bayonne at 13 o'clock. The countryside I have crossed is very barren and consists almost entirely of sand. There was only the odd small field sown with some grain. They say that all the sandy soil flows in from the sea to Dax, where the ground suddenly becomes more fertile. Dax is renowned for a hot spring. Bayonne is a fort above the beach. Desiring to see the small town of San Sebastián, I bought a ticket for that destination, plus a ticket for the journey from Bayonne to Madrid, for which I paid 140½ franc, because I was told I would find no ticket in San Sebastián. I can't tell you how happy and pleased I am to be on Spanish soil, surrounded by amicable people who are ready to help foreigners as much as possible. I have never seen anything as pleasant as the faces of the servants and the mistress at the inn where I stay. They are very attentive and polite and I'm in love with all that I've seen here so far. The streets are narrow, since the city is a fortress, surrounded by moats and fortifications. Outside the city are many fig and fruit trees. I saw both blossoms and ripe fruit on many apple and pear trees. When we crossed the boulevard outside the city, a balloonist had just taken off with his balloon. The man had attached himself to the lines under the balloon and was turning in circles. The balloon did not ascend very high, that would have been dangerous because of the nearness of the sea. Youth of both sexes danced and seemed to be enjoying life. I cannot visit San Sebastián without seeing Mrs Francisca de Santruntegui and inquire about her sons, who I got to know previously in California, and therefore I had a boy lead me to the house. The lady lives on the top floor of a house on Calle Mayor. Almost all the houses here appear

Figure 11: Madrid ca. 1854

the same in height and construction. I expected to come across a lot of depravity, but I was mistaken. [3] The magistrate and the municipality are very strict and evict any woman who practices prostitution and causes scandals. Thus, there are no prostitutes.

Wednesday 19/31 August in Madrid

After a very unpleasant two-night and two and a half-day journey, which I spent packed like a herring between the many other passengers who occupied the few seats in the coach, I finally arrived at the capital of Spain at nightfall. On the first day, I had a seat in the middle, between two others and it was so tight that we didn't know where to leave our feet and legs. On the other hand, I can't complain about the passengers, as they were amiable and polite. Although I was surprised at the conversation of the ladies, who talked about the most indecent things and used such words that I was embarrassed to hear them say. The plain we traversed is of very poor quality; everything is light clay, sand or rocks that often reach to great heights. The track currently being laid from the border to Madrid is facing great difficulties, because wherever a passage must be made there are rocks, mountains to be carved through and chasms to be filled or spanned by bridges or overpasses. All the villages we passed through were well built and you saw no trace of the very poor construction of our [the Russian] villages and provincial towns.

Thursday 20/1 September

Today I wrote to Mr Doré Wunderly, Calle de Rivoli in Paris:
Dear Sir and friend!

You were so good to have promised to send me the letters which were entrusted to Mr W[eissweiller] and B[auer][192] and which would arrive for me at the inn at the Louvre in Paris. I would be very obliged if I may inquire if my luggage has arrived, or if a message has been received that the luggage has arrived at customs. Please ask if they would be so kind to hold the luggage until my return.

The historic city of Madrid pleases me very much. A trading city it is not and it will be hard for you to believe that there isn't a single money changer here. But I have here a letter of credit and suffer no shortage of money. I was less lucky during the journey because I had believed that I could exchange Russian gold in San Sebastián. Not wanting to lose too much by first exchanging Russian gold to French gold and then again to Spanish coins, I had not changed more currency in Bayonne than needed for the seat on the coach. In San Sebastián I spent my last maravedis and despite the 30 pieces of Russian gold I had in my purse, I had to fast from San Sebastián to Madrid, a journey of two nights and two and a half days. Tormented by hunger, I offered several times to swap the imperial money for 1 duro, even though they are worth $4¼, but no one would accept them and the answer was invariably: never have I seen such coins, so they must be false and I won't give a penny for them.

[4] It is said that Madrid is very lovely in winter, after everyone returns from the country and coastal resorts, but now it offers a deserted and abandoned sight and the heat of autumn makes the stay here unpleasant for those who have enough resources to enjoy the fresh sea air. Today I traversed different parts of the city and I savored the faces of the female sex. There is nothing so gracious as a Spanish woman. They are generally dressed in black and have a veil or mantilla covering their heads. Their complexion clearly shows that they were born under the blazing sun of the South. Generally, the women here are large and they have much larger breasts than in our country. I would like to find a girl that would accompany me on my trip to the South and I would pay the person well who could recommend one to me. But, even as much as I travel, never would I dare enter a house of easy virtue, especially for fear, because of using the necessary condoms, that my penis would not remain stiff. I would have more easily found what I'm looking for if I had rented private rooms and advertised that I need a maid. Many would offer their services and I could choose the one that I liked the best. But I would have to stay longer than I wish and therefore, this time I must deny myself this very sweet undertaking.

Tomorrow morning at 11 o'clock I will go to the Russian Embassy to obtain authorization from them to see the royal sites within the city for free. After that, I will request some money at Hy P.'s bank.

The structure of the houses here is like those in Paris, only here they are not as sturdy, which one can easily excuse because of the great heat. The hostels cannot be compared with those in other parts of Europe. They do not even reach the level of the worst in Paris. Nevertheless, I am satisfied, because the thought that I have experienced much worse in my life comforts me.

192 In 1855 Daniel Bernhardt Weisweiller (1814-1892) along with Ignacio Salomón Bauer (1827-1895) founded the bank Weisweiler Bauer & Cia. This was the Spanish branch of the Rothschild Bank.

I was very wrong and have been misled about the Spanish nature, which I suspected to be too proud to welcome foreigners to their inner circles, or even to associate with. On the contrary, everyone here treats me amicably and welcomes me in a delightful way. I can never cease to thank God for the opportunity He has given me to travel.

Here in Madrid they have copied the French Emperor's example and have renovated and made improvements everywhere. For example, on the plaza in front of the inn where I stay. They have torn down many houses and built others to broaden the plaza and make it more symmetric.

[5] Today I told my Spanish teacher the many details of my ascent of the Vesuvius volcano. From my windows, I could see the volcano and the spectacle offered at sunrise was fantastic because, although the Mount Vesuvius eruptions stopped 6 months ago, the lava had not yet cooled and the mountain seemed surrounded by slow flowing rivers and streams of flames. On the third day after my arrival in Naples (having crossed through Russia, Sweden, Denmark, Germany, Switzerland, Lombardy, Venice, Sardinia, Romania, etc.) I traveled by train to the station before Pompeo, of which I have forgotten the name. Once there I hired a guide and a horse, a real Rozinante[193], which would have been quite pleased if I had carried it on my back. In addition, I was accompanied by two men, dressed as men of good stature, who had a litter with them. I suspected that they were only accompanying me to help me climb the volcano and since I had heard that even women could climb up without help, I preferred to climb alone and I explained to them that I had no need of them. The road led us directly through vineyards and orchards planted with fig trees and it was so haphazardly laid with stones that we could hardly make headway. When I questioned why a paved road hadn't been laid, the guide replied that the volcano's eruptions and the frequently occurring storms ruin the pavement. After passing the estates we wandered through huge fields of cooled lava, then a few streams with glowing lava. Nearby were a few men whose only activity was to dip a little of the glowing lava out of the streams with long sticks and then make it into the form of a [6] garland, in which they laid a copper coin, which they then offered to travelers to earn a douceur. We passed the hermit[194] "Tolerably high up the mountain, a full league and a half from the summit of the pyramid which is properly called Vesuvius, there is a hermit, who lives on a hill, by its side, which is called Somma."and after that it was not much further to the base of the cone, but the road is exhausting and painful, because one constantly has to cross the ashes discharged centuries earlier. There we left the horse and started the climb, which was extremely difficult, for the slope is very steep and for a 1000 feet is covered with ash in which you sink up to your knees and then you climb up over the pieces of lava that built the cone. The lava is slippery under your feet and one misstep would cause you to fall on your back. Finally, we came to the top of the cone and were duly rewarded for the toil and fatigue by the vast

193 Don Quixote's horse, an old bag of bones.

194 Stolberg, Frederic Leopold 1747 Travels through Germany, Switzerland, Italy, and Sicily II. London. 16 "Tolerably high up the mountain, a full league and a half from the summit of the pyramid which is properly called Vesuvius, there is a hermit, who lives on a hill, by its side, which is called Somma."

and magnificent view offered us in all directions. Pompeo seemed to lie at our feet. The smoke did not bother us because the wind blew it in the direction of the island of Capri. The top of the mountain is covered with fountains of scorching water, often covered with a thin layer of sulphur, and you need to be very careful if you do not want to graze your shins. We approached the crater, which belched thick smoke. But the crater didn't seem large enough to be worthy of Vesuvius. I asked the guide where the other was, because I reckoned that there must be two. He contested this and insisted that he would show me nothing more; I threatened to pay him nothing at all if he would not take me on a tour of all the craters and, by promising an addition to the agreed price, I convinced him to lead me. However, without warning, we found ourselves surrounded by the smoke. In vain we tried to get out; the smoke was so thick that we couldn't see our hands in front of our faces; we were lost and crawled forward on hands and feet. We wandered for two hours among the thin layers of sulphur that separated us from the fountains of hot water boiling beneath us and waiting to swallow us up. [7] The suffocating sulphuric smoke interfered with breathing and at times we thought we would choke. Finally the volcano took pity on us and answered our pleas. We managed to stay together by listening for our voices. Suddenly the guide called out: "We're saved" and we came out of the smoke. Three times we approached the craters in this way, whose chasm we found by touch. We could clearly hear how the smithies below us beat the iron on the anvil with their giant hammers, how the volcano fanned the fire with its bellows. We found our boy back, in tears because we had disappeared and crying because he thought he had gone to all the trouble of hauling the food up the mountain for nothing. He jumped to cook eggs in one of the fountains of boiling water and they were done in an instant. I haggled vigorously with him about the price of lunch and got it finally done for 5 Carolingians, for he had asked eight times as much at the start. I didn't know the strength of the wine and, because I was so happy that I was saved, I drank too much. As I descended the steep slope, I lost my balance and rolled down head over heels amongst the pieces of lava and ash. I could have broken my neck a thousand times, but God protected me and, except for a small wound to my thumb, I had no injuries. I crossed through Sicily and into Messina after which I embarked in the direction of Alexandria.

Madrid Friday 21/2 September

I did not like the room I had in the House of Viscayna [Biscay] and I asked if they could give me another. They could and I switched rooms. The reason I wanted to move was that my previous room was too dark. Although I often felt tickling, I don't believe that there were bedbugs, fleas or lice, in any case so far I have not managed to spot any of these parasites. I went to the Russian Embassy today, near the gateway through which we came coming from Bayonne. [8] However, I made an unnecessary trip, because, as the doorwoman told me, the Chancellery was near the Plaza del Oriente. I hurried there. There they informed me that having my visa and passport signed was not necessary, as I was already endorsed for Spain and Portugal. I asked what was necessary to gain access to the royal residencies and they said that I just had to show my passport. Then I went to the Office of

Weissweiler and Bauer and requested 100 duros from my letter of credit account. They answered that the credit was in francs and that they could only pay me in that currency. They subtracted their commission and paid me 1880r or $ 94 in paper money. I had asked for gold, but that they did not have and they said I could exchange the paper currency into gold anywhere, for a ½ % surtax. I told them that I had almost died on my route from Bayonne to here, even though I had 30 pieces of Russian gold in my pocket, which I had offered for one duro per piece, even though they were worth $ 4¼. That made them laugh. I left and on the street met by chance the lawyer Mariano Demetrio Ortiz, in whose company I had made the journey from San Sebastián to here. I went with him to a café and he promised to introduce me to a very reliable private house the following night. I would no doubt find the possibility to establish contacts with a maiden who would agree to accompany me to the South of Spain. Today, even in the shade, it is particularly hot and anyone who is not forced to go outside, stays indoors.

We had hardly dropped anchor in the port of Alexandria when a small boat made its way to us with many Arabs shouting: "gentlemen, a token?". That was the first and the last word I heard in the East. Often, as we sailed the Nile with favorable wind and great speed, they cried: "give us a token," and as we had to dock, it was hard not to give them what they asked. They often jumped into the water and swam to our boat, where they remained until they got the desired gift. [9] The Shirocco [Sirocco] wind, which in Italy is damp and quickly destroys the monuments of antiquity, is very dry in Africa, for it only takes on water above sea. It never rains in Egypt and the dryness of the climate conserves the monuments such that, for example, the paintings and drawings made 4000 years ago look as fresh as if they were made yesterday. The pains to which I had voluntarily subjected myself because of a certain Jewish porter's scam, made me do my best to learn the Arabic language and I did manage to speak it within a few weeks. The Arabian homes are poorly built and consist of four palm tree stumps bored into the ground and then filled with mud from the Nile. The roof is made of palm leaves placed on top of improvised walls. The doors are so low that you must crawl on your knees to go inside. Inside, the furniture consists of a bulrush mat, upon which the whole family sleeps. The clothing consists of dirty rags. Until the age of six, the women walk around totally naked and they wear just a narrow belt covering their pussy. On one occasion, I heard a loud shouting coming from a house in Corusco. I went inside to see what was the cause of so much noise. Inside I saw two men who had a newly born child of the female sex on the table. The mother, armed with needle and thread, was busy stitching shut the little girl's pussy. I asked what the reason was for such a painful surgery. They said that it was to prevent prostitution and to preserve virginity until marriage. Then they explained to me that after surgery the girl would be placed on her stomach to sleep until the sewn part had healed.

Tonight, I made my rounds on the beautiful promenade called 'el Prado' [10] and which was full of people of all classes and I was filled with joy, because I can think of nothing finer than walking amongst the beautiful Spanish women. When we feel happy, time flies and I felt that the afternoon went quickly. Suddenly my happiness changed into a nightmare, because I remembered the family F[errier] with whom I had partnered in the East and whom I had promised to visit in Canada, but circumstances beyond human efforts prevented that. I decided to send the father the following letter:

When I took leave of you in Jerusalem and you informed me how sorry you were that, because I already had other commitments, I could not journey with you through Syria, my answer was: "Divine Providence" has made it so, and later we shall see that it was better that way. To tell the truth, I had only just arrived in Bethlehem, when the memories of my former life provided powerful reasons not to settle in Canada. Human nature has made us so that we feel relief and consolation when we can talk to others about the things that weigh heavily on our hearts. I hurried to confess to Mr Smittan the reasons that, as I see it, prevented my settling in Canada, and the reason why it tempts me and makes me hope to be the happiest among mortals. Amid the desert rocks and in the enormous heat on the shores of the Dead Sea, I debated this matter with S[mittan] everywhere; this was the only thing I could think about while awake and about which I dreamt when sleeping. Finally, we both decided that the man [11] who wants to be worthy of the road that God has chosen for him in this world should give up the joy of his life rather than to besmirch his conscience or be distracted from his highest principles. If you had not rejected my company for the trip to Jerusalem when I offered it you in Cairo, or if, after testing my pride so profoundly in Cairo, you had said a single word in Gaza, where I repeated to you three times, in the purest of English, that I had commitments only until Jerusalem and that afterwards I would be free to join any other company I chose, then I would have gladly accompanied you, not only throughout Syria, but undoubtedly also to Canada, because the sincere friendship to you and your family would have made it impossible for me to separate from you. Back in Jerusalem, I was determined to prefer losing my life to going to Canada. A deep sadness and a grief as I've never felt before gripped me. In such a mental state, I would have been tedious company for my companions. I split off from them and toured Syria with a dragoman [Turkish servant] and two domestics and it was my joy to pitch my tent at the same spot where, as a few natives indicated, you and yours had stood a few days earlier. In Nablus, I slept in the house belonging to a certain Jew; the same where you found refuge after you arrived there soaking wet. I came across the last traces of you in Tiberius. Since I'm used to being in the wide world and living in the company of people, I was very bothered by the loneliness and when I could bear it no longer, I decided to go straight [12] on to Greece, without having visited Lebanon and Damascus. With this resolve, I arrived in Sidon and went to the French Consulate to inquire about the departure of French steamships. As luck would have it, I came across my good friends there, the two Italian counts[195] in whose company I had made the trip from Cairo to Jerusalem. From them I heard that you had left for Constantinople the night before. In Beirut, I met your acquaintance, the priest Taylor, who told me that you planned to go directly from Constantinople to Greece, because Miss Mary wanted very much to see that country. So, if I had gone to Greece, I inevitably

195 The brothers Carlo and Giulio Bassi.

would have come across you, and I didn't want to see you, as it would cost me my life if I were to journey with you through Greece. So, I decided to strike out in the opposite direction and I went to the mountains of Lebanon. I traveled through the perennial snow to see the living cedars of Salomon, visited Baʿalbek and Damascus and boarded ship in Beirut two weeks after you left, purposely staying on the Greek Islands and in Smyrna for a few days. Trembling over my whole body for fear that I would run into you, I finally arrived in Athens. It would be impossible for you to even have the faintest idea how much I have suffered and the toll it takes to be separated from you; you whom I love and value above all in this world. My anguish was indescribable. The lasting grief affected me so deeply that I became seriously ill in Athens; a nervous fever kept me bed-ridden, little by little I lost all my strength and was so weak that I couldn't move a finger. I felt powerless for hours at a time and I expected death at any moment. While I was in this desperate state, they brought me letters from my banker in St Petersburg. These letters were in Swedish, no one could read them for me at the inn I was staying at and I was too [13] weak to read them myself. The doctor went to the Swedish Embassy and asked the Secretary of the Embassy to read them. He came to my bed and read them to me. The drift of these letters raised my bile, as the banker told me of a certain individual called Solovieff, who for 18 months has owed me £17,000. When he proved unable to pay, I had the decency to allow him to pay only a fifth in cash and the rest in bills of exchange every one, two, three and four years. He not only had not paid the first instalment, he had also filed a complaint against me that attacked my honor. Filled with rage, I asked to have a wagon immediately take me to the French steamship of the imperial courier service, which departed that day in the direction of Constantinople. At first the doctor opposed, but later he agreed. So, I left behind a reimbursement for his services and had me taken to Piraeus, the port of Athens, and placed on the bridge of the steamship. The fresh sea air cured me quickly and after a few days in Constantinople I was walking again. I went to the mouth of the Danube River and upstream up to Pest, across Germany to Stettin, from where the Russian steamship brought me to St Petersburg in three days. There was no need to return there to receive payment, because in our country, as everywhere, there is a law that requires every bill of exchange of which the signatory has not stated within three days of signing that he hasn't received payment, must be paid. I only returned to salvage my honor and having arrived in St Petersburg, I deposited all of S[olovieff]'s letters of exchange at the trade court. I entrusted the simple defense to the most skilled lawyer and left for Paris, because I was afraid my fever-weakened health would not endure the anger that the trial would bring. [14] I stayed in Paris for 12 days and attended the Napoleonic festivals. Then I went by way of Bordeaux and the Pyrenees to Spain, in whose capital I have been since a few days. I will be here a few more days and then go on to visit Granada, Malaga, Seville and Cádiz. After that I will visit Portugal and I'm thinking of spending the winter alternately in Paris, Rome and Naples.

On the eve of my departure from Jerusalem, I was delighted to have visited the tombs of the Kings with you; the square where the tree grew from which trunk the Holy Cross was made, and lastly the quarries from which the stones were cut for the construction of the Temple of Solomon. I recited to you the famous speech that the unfortunate Kossuth had held at the Capitol of Washington in July 1851 and in which he says: "The palm-tree grows best beneath a ponderous weight ...". I could not stop myself and roared with laughter upon hearing these strange words. But the hard and bitter experiences have since taught me how true the meaning of these words is.

In the joy and happiness of the heart we often forget to thank God for his favors; we never forget it in the sorrow and sadness. On the contrary, sadness of the heart makes man lay claim to the highest principles of honesty and virtue with a fiery zeal that fulfills and elevates the soul.

The faces of you and your family will always be present in my soul and it shall be my pride in this life and my dear and faithful talisman in the hour of temptation.

I would not trade the memory of you for all the riches in the world. How comforting and sublime, in the sadness and sorrow of the heart, is the faith in another, happier world. How heavenly for me is the hope that I will see you again in that world and then perhaps shall live with you for an eternity in the harmony of a close friendship in which our lives are just a [15] moment in comparison. Take heart, miserable heart! Take heart in faith! Asking you to carry my warmest greetings to your wife, your daughter and your son and repeating that I shall never forget to include you and your family in my prayers to God, with feelings of deep respect and love, I remain, your friend Heinrich Schliemann.

In earlier times, money was the only subject of my thoughts and Mammon had never had a stronger supporter than I. Now money finds me completely indifferent to it; as I am not prone to excesses, I don't think I'll ever deplete my income, or even spend half. In Gaza, you told me that in your city one could easily earn 10% when buying assets. I would be happy to send you money for investments, but then I would have to start a correspondence with you and I feel that I don't have the strength for that, because every letter from you would reopen the wounds in my heart. I hope never to return to St Petersburg, because it is too cold and the 3½ weeks I spent there in July, I sat in my house wrapped in my fur cloak. But my name is well known there and if you simply write to H[einrich] S[chliemann] in S[t]. P[eters]b[ur]g, your letter be sent immediately and it will always reach me by mail, because my agents always know where my letters will find me.

After your departure, a newspaper cutting was found in your room, which said the following: "the venerable T[aylor] honored us with his visit last week, but, like a comet, he departed, leaving behind a giant tail of light."

[16] Madrid Saturday 22/3 September

I have the habit of washing my whole body with cold water in the morning. This morning, no sooner had I done this than the teacher knocked on the door. I said: "Come in," and the big man appeared. We went for a stroll in Retiro park, where we arrived after having crossed el Prado boulevard. I think it takes a lot to keep this garden in good condition and regular watering of the plants, because the soil is sandy and the hot sun burns everything. The garden is famous for the enormous number of beautiful flowers and for a few fairly large ponds, deep and beautiful, where ducks swim. Accustomed as they are to being fed, they rush toward us, probably hoping to get thrown some pieces of bread.

Since the island of Cuba contributes much to the prosperity of Spain, it would seem to me that the latter would do everything to support that country. Everyone knows that the present President of the United States was chosen for no other reason than that he is an advocate of the annexation of Cuba. If Buchanan had been President during the Crimean War, when England and France were too busy to think of intervention in a war in the West Indies, without the slightest doubt, he would have attempted to take over Cuba. So, because I love Spain and its courageous inhabitants, I hope that they forgive me if I, concerned about the future prosperity of Cuba that is so important to them, recommend sending an army as big as they can muster, because otherwise the Americans may take advantage of the circumstances. [17] Looking at the geographical map and gazing through the hieroscope[196] at the peoples, I predict that great changes will take place in the next five years. The Greek poet Homer, who lived 1000 years before the beginning of the Christian era, says: ἔσσεται ἦμαρ ὅτ ἄν ποτ᾽ λώλῃ Ἴλιος ἱρὴ αἱ λαὸς ἐϋμμελί Πριάμοιο, which means: "the day will come when the holy Ilios will perish and the people of the well-dressed Priam". In the same way I say: "the day will come, and that day is not far off, when the Russian fleet, together with the French fleet, will destroy the so far invincible English fleet, and when proud England will be reduced to an even weaker state than the tiny kingdom of Holland. Canada, ashamed to belong to Great Britain, will join the sovereign states of North America and sparkle under the flag of stars." American filibustering [197] will then conquer Jamaica, with hardly any opposition from the lazy Negroes. Cuba, the Pearl of the Antilles, will be made an American territory and subsequently war will break out between Russia, France and England. The bloodshed will be huge, because the Spaniards will heroically defend their island. As war against England breaks out, the English will call on the other countries for help; to no avail, no one will help them because everyone hates England and will rejoice in her fall. Gibraltar will then be returned to Spain, to which, per law of nature, it belongs; Russia, Moldova and Walachia will take European Turkey, except for Albania; Macedonia and Thessalonica will be returned to Greece. France will seize the Ionian Islands and, together with Russia, proportion Asia Minor, Syria and Palestine to spread the civilization [18] that has been oppressed by the English barbarity. The East Indies will be fraternally divided between France and Russia; Australia will proclaim itself a republic. Victor Emmanuel, King of Sardinia and Lombardy, will remain the vassal of Napoleon. To limit the power, the Duchies of Tuscany, Modena, Parma and Romania [part of the Papal State] will be fused into a separate Duchy. The spiritual rule of the priests will remain in Rome for the time being and the Pope will not be exiled for another five years. Taking hieroscopy a bit further and adjusting for the events in the years 1865 – 1870, I see France taking all of Italy, except for Sardinia, which it leaves to

196 195 Hierscopy: oracle, divination by inspection of entrails of victims offered in sacrifice.

197 196 Originally this term was used for the intervention in the politics of independent states. It is now better known by its more recent meaning, a prolonged speech as a delay tactic.

its vassal. You must excuse me my hate-inspired words about the English. My hate is limited to the English and does not extend to the humble Scots, whose sincerity, faith, affability and natural intelligence have always very much pleased me. As we have the habit to paint a picture in our imagination of unknown things that interest us, so I have a picture of Montreal. It would be a beautiful and clean city, very similar to Philadelphia (although much smaller), with an excellent harbor and a population with a lot of community spirit. This combines with the American spirit of enterprise and the thoughtfulness and virtue of the Scots. At the round table in Jerusalem I gave you a fantastic example of this: the villagers of Kirkaldy, one of whom became a Turkish wazir [high-ranking political advisor] and the other a Russian field marshal in 1739.[198]

[19] 27 May

Unfortunately, Spain has always been visited by foreigners who, since they knew not a word of the language, could not be welcomed with conversation and were irritated by this. They could not, of course, gather information and left Spain dumber than they arrived. Nevertheless, they set about writing books about Spain, books full of the most terrible lies. As if it was perfectly normal they told of how they were attacked and robbed in the streets of Madrid. That is why everyone has very erroneous ideas about Spain. I can assure you that never in my life have I been better received and the only thing I did was treat the people with a little pride and simultaneously, a loyal sincerity. A friend of mine, who has lived here for some time, assures me that he was overwhelmed with invitations here and that he experienced so much hospitality, more than he had even found while in Russia. The Spanish nature is a little proud, but this pride is commendable, for it comes not from vanity but from the knowledge of one's own value as a human being, and one feels indignation and contempt for those who lower and humiliate themselves to acquire passing advantages. A natural consequence of the hot climate is that the Spaniard is more vivacious than the Frenchman, but his vibrancy, lined with his pride, bestows an indescribable grace to all the movements of his body. This seemingly natural grace is realized by a nature that nurtures contrasts and that (by the mixing of the blood of Saracens, Romans, Carthaginians and of those from other countries who have inhabited this land one after another) brings out the wonders of beauty in both sexes. The female beauty surpasses anything I've

---

198 Littell's Spirit of the Magazines and Annuals 2, 1838, 209. "The story, we should suppose, is also well known, of a certain vizier to the sultan of Constantinople, having been a Scotsman from Kirkaldy. It is as follows: at the conclusion of a war between the Russians and Turks, before the treaty of peace was concluded, there was occasion for a conference between the Russian general, who was Field-Marshal Keith, and the grand vizier, to settle some preliminary articles. When the conference was at an end, they arose to separate; the marshal made his bow with hat in hand, and the vizier his salaam with turban on his head: but when these ceremonies of taking leave were over, the vizier turned suddenly, and coming up to the marshal, took him freely by the hand, and, in the broadest Scotch dialect, spoken by the lowest and most illiterate of our countrymen, declared warmly, that it made him 'unco happy, now he was see far frae hame, to meet a countryman in his exalted station.' Marshal Keith was astonished: but the vizier replied, 'My father was bellman of Kirkaldy, in Fife, and I remember to have seen you, sir, and your brother, often occasionally passing.'

ever seen so far. A natural consequence of the hot climate is that the women here mature and wilt faster than in more northern regions, but what always remains is the indescribable grace of the movements of the body. [20] Because their great vibrancy blends with their dignity, they speak with unfailing genuineness. The national pride and patriotism necessitate everyone here to care about the welfare of the country. Even the humblest maid knows all the details and the discussions in the Cortes [House of Parliament] and there is not a coachman who has not, seated on the bench of his carriage, leaning into the light of the lanterns at night, devoured the debates of the Cortes and all the other news that the newspapers offer. At an elderly age they add a wealth of experience to all this so that one could very well fall in love with a Spanish woman of 65 years, or at the least, show her nothing but deep reverence and respect.

The Castilian language is so melodious, rich and harmonious that all the other 15 languages I speak are ugly in comparison. Convinced of the beauty of their native tongue and too busy with all the other pursuits of knowledge that could contribute to the improvement of the state of the motherland, the Spaniards study no other languages. Therefore it is necessary that every foreigner who wants to enjoy his travels in Spain, and who wants to enrich himself with important information, must above all have perfect command of the language. The hot sky of Spain is the reason that the passions here are more fiery and fervent than in the more northern countries, and the laws of nature do not seem to be in harmony with the law of man that stipulates monogamy. It can also not be denied that

Figure 12: Spanish woman, 1854 – by Jean Désiré Gustav Courbet (1819-1877). Atkinson Art Gallery Collextion – Southport

this latter law, although formally maintained in name, in truth is not, and that Spaniards find that they by nature have the right to emancipate themselves of the said law. Therefore, there is more freedom here between both sexes than in any other country; often the conversations of even young ladies and women of the higher classes revolve around indecent affairs, and they use such words that even my ill-mannered [Russian] maids would be too ashamed to whisper in each other's ears. A similar emancipation of customs also reigned in ancient Rome and [21] ancient Greece, as is evident through the writings from ancient times and by the paintings and emblems that you [Ferrier] will have seen in certain homes in Pompeii near Naples. In countries where Islam allows polygamy, the passions of men are permitted to be satisfied; the laws of nature are in accordance with those of the people and that is why there is almost no prostitution.

I rang the bell and when the attendant came, I said: "bring me a coffee". But I told him: "I do not want cold coffee like I got last time. If the coffee is not very hot, I don't want it; send it back and ask another. First you need to pour a little boiling water in the cup to heat it up. Also, bring me bread with dairy butter and salt and mustard, because I do not wish to have lunch." But, since I am thinking of going to the bullfights today and because they last until 6 o'clock in the afternoon, perhaps it's better if I do lunch.

Madrid Monday 24/5 September

Yesterday I invited an acquaintance to dine with me. He's a thin little man and I can hardy imagine that he has, as he says, three women, two steady and one for fun. After having eaten a herbal soup and bread, good fat beef with vegetables, bacon with lima beans, sausages with green peas, tomatoes, beans, roast lamb, roast veal, pudding, ice creams/sorbets with puff pastries and cookies and for dessert cherries, pears, apricots, figs and sour cherries, we left and went to a café. They serve coffee in a crude way; they fill the cup in such a way that they spill on the table and the parishioners' [guests'?] clothing, and they don't serve a glass of cognac at all. Moreover, the coffee is bitter and distasteful. Then we took a long walk along the boulevard de la Castellana [Paseo de la Castellana] and we went to an open-air dance party. [22] It seemed that all of Madrid's seamstresses of common origins were gathered here. The dances seemed different from ours, because here the woman lays her head on the man's chest and it does seem as if he fucks her as they dance. One can have no better date with a woman than at these dance events. There were many more women than men and since everyone had picked out the beautiful ones, all the ugly ones remained neglected. Two bands played music fairly well. The garden was well lit with torches of all colors. After having enjoyed watching this dance party for half an hour, we went to the Prado and hardly had we left or we saw the difference between the type of people dancing and those on the street. The Prado was crowded with people strolling. At the end of the boulevard a charlatan [street vendor?] had set up a wooden pavilion and by the beat of a drum a colored man and a young boy showed the public the wonders of the art of this charlatan. Both lacked front teeth and they took artificial dentures that had been prepared beforehand, placed them in their mouths, showed how sturdy they were and started to chew large pieces of bread that they stuffed into their mouths. At the same time, they passed out papers on which the details had been written.

I did not manage to see the Royal Palace today because the government is seated there all summer. But they handed me papers for entrance to the Royal Armory, where they store their antique armor. Although it was closed today, I gained access by showing my passport. When viewing the heavy armor, the degeneration of the human species pained me, because nowadays it is only with great effort that someone could stand up in the armor that the ancients wore for whole days. [23] Nevertheless, much depends on exercise and if, rather than cultivating the mind, we would submit our body from childhood to gymnastic exercises, perhaps we would be just as strong as our forefathers. In the Armory are a large number of sabres, swords, hunting rifles, spurs and other materials. Visitors are warned everywhere that they are not allowed to touch the objects. Afterwards I went to the Royal Stables, where an immense collection of beautiful horses of all breeds could be found. There is an infirmary, a workshop etc. I'm afraid to burden my stomach, as I believe that my sadness and melancholy only come from my abundance of blood, as a normal consequence of the little movement that I do and the ample amount of food I'm used to consuming.

It is with great pleasure that I will always remember my trip to Jerusalem, although I probably won't make this journey again in this lifetime. Life in the tents in the desert was certainly extremely one-sided, and I often had to subject myself to of all kinds of great hardships. However, if I now look back it seems like I see everything through a rosy fog, because it leaves behind many glistening points on the misty horizon of memory. The road from Jaffa to Jerusalem is one of the worst there are, although it is frequented annually by thousands of pilgrims. It is said that the road used to be better and that it was destroyed by the Arabs to stop the advance of the Egyptian army. Everything is cluttered with stones of numerous sizes, so that even at a walking pace one risks breaking one's neck. Jerusalem comes into view only a few minutes before arrival. My joy was boundless when I finally saw the object of my journey before me; great was my disappointment when I entered and traversed the narrow streets that were so slick that you ran a great risk of slipping while on horseback. The temple or Church of the Holy Sepulchre is in poor condition, [24] as the roof of the dome has gradually decayed, so now the rain falls without mercy and pity on the heads of the righteous. The rivalry and the irreconcilable mutual hatred between the three dominant religions is the reason that the Holy Church threatens to become a ruin. If the repairs would cost millions, any of the sects would come running to do what was needed, but the one won't grant the other the honor to do the repairs, and since there is neither consensus nor agreement among them, the Church will be a pitiful pile of rubble within a few years. The traveler enters the Holy City turns into outrage as he looks at it carefully. It contains both the Holy Sepulchre and Mount Golgotha, where our Lord was crucified, yet it is impossible that both are found in the middle of the city, which could not have been larger then than it is today, because it is surrounded by high and sheer rocks. Nor is it possible that the grave be found so close to the place of the crucifixion. Jewish law forbids burial of the dead within 100 feet from the city. In the middle of the church is a chapel which encloses the grave. The chapel consists of two parts. In the first part is the post of the pillar to which Jesus was tied when the crown of thorns was placed on his head and when he was whipped. In the second part is the grave, but its marble seems too white for me to be able to believe that this could be the real coffin. In the second part is always a monk who, when the faithful come to offer their prayers to God, gives them one of the flowers from the glass vases and anoints them with eau de cologne, while at the same time he extends his hand to accept a favor. This little business at the holiest of places I find most annoying.

[25] Mr Oriz came calling at 4 o'clock and together we went to the bullfights. The building where the fights take place consists of a large amphitheater that holds 17,000 people. It was an extraordinary spectacle to see the desire with which the people hastened to the bullfight. On everyone's face was the fear of being too late. From the Puerta del Sol up to the arena, mounted police were stationed everywhere to keep the order at the carriages. Those arriving had to join the queue and go forward in such a way that the first in line were the first to drop off their passengers. The empty carriages rushed back as soon as possible to pick up other passengers. The coaches were stormed and besieged. We had good places that were called tabloncillos [wooden benches], first class. It is very good that the money earned from the fights goes to the benefit of the poor and the sick. This enhances the pleasure that the Spanish have in bullfighting. The poor still prefer hunger or would pawn their last pot at the 'mountain of mercy' [pawn shop] to get the necessary money together to see the bullfights. The crowd was huge; it was an extraordinary sight to see so many people gathered together. As the music blared, the doors of the arena opened and the guards on horseback came out, followed by the toreros and a team of mules that was used later to drag away the dead horses and bulls. The men bowed low to the president of the bullfight and departed; the toreros took up their capes and the judge gave the stable's key; the gate was opened and a ferocious bull came out and charged the toreros, who held their capes in front of the bull's eyes and then skillfully escaped by jumping over the parapet. [26] Three bullfighters remained mounted, equipped with long lances topped by small spikes. As the bulls themselves charged them, sinking their gigantic horns into the bodies of the horses, they would spear them with their spikes, but they could only cause minor injuries. At nearly every attack of the bulls on the horses, they would fall with their riders, whose legs were cased in iron so they were not injured by the fall. As they lay on the ground their position was very dangerous, but the toreros on foot were there straightaway and saved them by brandishing the capes before the bulls' eyes. No sooner had a bull gored a horse's belly than the guts would spill out – often there was just one intestine that would be caught up under the hind legs of the horse. Thus, by stepping on it, the horse itself would pull out the rest. The bullfighters don't take heed of the poor animals and let them gallop until they drop dead.

Figure 13: Bullfight in 1850 in with the famous Matador Francisco Montes Reina (1804-1851) was injured

Often the bulls would leap the fence, but no one was injured. Then came the bullfighters with lances that were wrapped in paper in the shape of a feather and they pierced these with great skill into the bull's neck. The toreros were cheered when the lances were well placed; booed if they were not and fell after they were placed. Several times the wrappings of the lances contained sticks of fireworks that would ignite and surround the flabbergasted bull with sparks. [27] Then one of the bullfighters, armed with his sword in one hand and his cape in the other, which he always held in front the bull's eyes, would try to find a favorable moment to plunge the sword between the bones of the bull's neck and make the kill. I felt very sorry for the horses that had to gallop around with half of their intestines hanging outside their belly. No sooner was a bull dead than the mule teams were hitched to the animals' necks and amid whiplashes, the carcasses were dragged out of sight. I find these performances too cruel and too violent and I think I must get used to finding them beautiful.

I'm thinking writing the following to Mr Fehleison from St Petersburg, currently staying in Le Havre.

My dear Sir!

No sooner had I taken my seat in the carriage to Madrid, than I understood that the Spaniards were still very backwards in comparison to other countries, because there we sat with six people in a space that could barely contain three. I really wished that my legs were made of rubber. I've been in the capital eight days and I must confess that even travelers from Paris can amuse themselves for some time because of the big differences there are in the customs, the way of life, the public entertainment – in short, in everything. The hot Spanish sun makes the blood fierier here than in other places. Everyone feels bullied by the unbreakable law that men may marry only one woman. Also, I think by what I see and hear, that not one husband in Madrid is faithful to his wife (and few wives faithful to their husbands). Prostitution here is even more common than in Paris, but here love is given on but a whim and is not for sale as it is over there. The truth is, the Castilian woman is too proud to sell [28] her love and she would certainly not lower herself to this unless forced by extreme necessity. I do not need to assure you that there is no permanency in the whims of the female Castilian and that their love spins like a weathervane. If I were to tell you that the women of San Sebastián were beautiful, then I can assure you now that nature has done much more its best here [Madrid] to produce miracles of beauty. The exalted Queen of Spain is a rare exception to the general rule, because she lives with her exalted husband in a union of the most ardent love, the proof of which is the gift she has given to him of eight children, all as beautiful as rosebuds. Their love of material things is most evident in the public entertainment. If the government were to proclaim today that there would be no more bullfights, a terrible revolution would break out in all of Spain; if the open-air dance festivals would close for a week, the population would hold them out in the open road. As I have said, the love of the material is the natural fancy of the people, with boisterous rowdy entertainment. Yesterday I was seated in the bullfighters' arena; the picadors mercilessly kicked their horses and made them gallop while their bellies had been pierced several times by the sharp horns of the bull, so that their intestines hung in large masses outside their abdomens, or were, bit by bit, pulled out by the horses themselves as they stepped on those intestines with their hind legs. Meanwhile the toreros pierced the neck of the bull with many lances that were filled with sticks that, when ignited, would explode and engulf the bull in a sea of flames so that it roared from pain and despair. I could not stop myself from crying out: "what a terrible spectacle". A lady who sat near me heard my words; blushing with rage she said: "What horrible bad taste you have that you're not moved and affected by the sight of this sublime spectacle." Really, she added, there is not one among the 17,000 visitors here [29] so crude as not to be electrified at such a heavenly sight. And there seemed no end to the happiness and joy that were expressed through constant applause and the shouting of "bravo". The cock fighting is no less terrible to watch. I do not think that Divine Providence has given man the right to torture those poor animals which, like us, feel pain. When the Spaniards mock us because we still have servitude in Russia, I answer: "In our country, only some of the villagers are indentured, and their servitude ends within a few months,[199] while your servitude will never stop, because you are slaves of your passions and this slavery suppresses you much more than the village boss's whip has ever done."

On the second day of Passover [Easter] in Jerusalem, it has always been the custom that the Holy Fire be shared amongst the faithful gathered in the church. It was said that the fire would suddenly emit from the Holy Sepulchre. The Greek patriarch, who believed this medieval custom unfit for our age of progress, had abolished this trickery. But the people loudly called for this custom to be resumed because, as they said, without it Easter is not the same. I took part in this trickery that usually costs many lives, because the masses are so large that many suffocate or are trampled. Thanks to the English consul we had access to one of the galleries from where we

199 Serfdom was abolished in Russia in 1861.

could see the spectacle. Many had spent the night in the church to obtain good spots or at least hope of a spot. [30] I leaned over the balustrade under the weight of two fat clerics who leaned over me. Others pushed me from both sides, so I was sweating like I was in a Russian bath. After I had endured this for three hours, they made it seem as if the fire had come from the grave, because one of the persons who had approached the opening in the chapel and had reached in with his wax candles suddenly pulled his hand out with the candle lit. He lit a few other people's candles, who in turn passed the flame to others, so in few moments the whole church was a sea of lights, while the cries of jubilation continued. Then the Turkish soldiers, who were stationed there by the church to maintain order, made way for the processions of the clergymen of the three dominant religions that made three rounds around the chapel, carrying sacred sculptures and crosses. I could not stop thinking what Christ would say if he would suddenly arise in the middle of such proceedings. The temple's courtyard is always full of merchants selling crosses, rings and all kinds of holy jewels of mother-of-pearl made in Bethlehem. Two days before my departure I saw a Jew with mother-of-pearl jewelry enter the room of an American traveler who was staying in the same hotel and shortly thereafter the woman of my heart also went inside with her father. I went in as well and after I had given the lady several figurines, I asked her to point out the most beautiful one for me. She chose three, of which one was a box which had on its cover a depiction of that stately entry of our Lord into Jerusalem, seated on a donkey. The other displayed a very nice illustration of the Lord's Supper and the third was also an event from sacred Scripture, but I do not recall which one. On the back of each of these jewels I wrote: "to the missus of the Honorable Mr J.F. With [31] compliments from H.S. and his best wishes for your everlasting wellbeing." Then I wrapped them separately and then all together in Greek newspapers. In the small box, I had slipped a note (folded many times over to make it fit) with the following contents:

My dear Madam!

I had already noticed on board the steamship that brought us from Messina to Alexandria, that you and your family had friendly feelings for me, which I met all the more enthusiastically when I saw that they are not based upon selfish interests. When I was poor I had no friend at all, and I am convinced that all those who became friends after the Crimean War, having increased my fortune six times in comparison to what I had upon my return from California in 1852, would leave me if I stopped treating them to dinners and champagne or passing on profitable business. I can't possibly explain to you how grateful I am for your selfless friendship. Since I have known you and your family, Canada seems a paradise to me and I would be the happiest man in the world if I could live in your vicinity there for the rest of my life. I promised you I would come to Canada this year, but I have been deceived a 1000 times over in my surest hope and I should not have promised you future things that depend on Divine Providence. Man's fragile fingers can neither stop nor slow down the turning of the wheel of human fate! If I am not destined to see you and your beloved family again, then I shall forever say farewell to you. Rest assured that I will never forget to include your and your family's wellbeing in my prayers to God and that, though I be in the middle of the hurricanes above the roaring oceans and in the enchanting monotony of the Nile Valley, on the icy shores of the Gulf of Finland and above the smiling waves of the Mediterranean, in the midst of the amusement of the fraternities in Paris, London, St Petersburg and Moscow and during the urgent pursuits of trade, [32] in perils and exhaustion, during work and adversity, never will a single moment pass in which my spirit won't be with you, whispering blessings to you, rejoicing in your luck and crying over your failures. With feelings of deep respect and reverence,

Sincerely yours,

Your friend.

My pride was wounded by your refusal to allow me join your party in Cairo, but so much worse in Gaza when you left me without inviting me to accompany you from Jerusalem to Beirut, even though I had said to your wife while in Cairo that my arrangement with the two Italian dukes was only to Jerusalem, where I hoped to join up with you, and although I told you twice in Gaza that I would be free to join any group whatsoever in Jerusalem, and the third time was when Mr Smyttan had invited me to accompany him to Petra, and that I had said that I would have to think about it. I had hoped at least upon my arrival in Jerusalem to find a note at the Hauser Inn, inviting me to accompany you on your journey through Syria. When I found nothing, I began to think that God was handing me a warning not to go with you. The 'Great Comet', which was observed late December 1858 in Constantinople, shall have long passed the Strait of Gibraltar. Its passage of the first Nile waterfalls, of Jerusalem, of Lebanon and Mount Sinai, will have contributed even more to its glory, and the Canadian astronomers will observe this comet in all its glory for many years to come.[200]

[33] Across from my windows in the Vizcaina [Biscay] Inn lives a gentleman whose maid, a loving little creature, let me know that she'd like to be fucked. Not knowing how to pass her a note, I shouted to her, as I close the curtains, that she should come to my room at 10 o'clock. I don't understand why she didn't come.

Madrid 26/7 September 1859.

Yesterday I went to the paintings [art] museum. The collection is perhaps the largest in the world and there are many masterpieces, from both the Spanish school and the Italian school. There are also a few rooms with sculptures that offer many fine pieces of old and modern art. Several copies of the kind of decorations on the walls of the Alhambra hang there. With their vibrant colors the decorations are a wonderful sight.

We went on the road and after one and a half hours we arrived in Bethlehem, where there are many monasteries and beautiful houses belonging to the bishops. Even the houses of the villagers look very clean and there are about 400 inhabitants.

200 Bond, G.P. 1862 Account of the Great Comet of 1858. Annals of the Astronomical Observatory of Harvard College III. 364 "The whole period of visibility of the Comet extended from June 2d, 1858, to March 4th, 1859, an interval of two hundred and seventy-five days. It was seen with the naked eye from Aug. 19th to Dec. 9th, one hundred and twelve days. The tail appeared first on Aug. 14th, 1858, and was in sight until Feb. 9th, 1859, or for one hundred and seventy-seven days."

The women are distinguished for their beauty. We entered the church and approached, with reverence, the manger where the son of God was born. The manger is hewn from a rock. We continued our journey and arrived in Hebron at noon, where to my surprise, the Jews spoke to me in Russian. We pitched our tents and went to the tree – a holm oak – called the Oak of Abraham, but it didn't seem that old to me. Of course, the bark was full of travelers' names who wanted to immortalize the moment. In Hebron is the mosque in which Abraham, Isaac and Jacob are buried. But since the Arabs very fanatic, I didn't dare to go inside. The next day we went to the camp of the Bedouin sheikh. The camp consisted of many pitiful tents. [34] Our unexpected arrival spread joy on all their faces. The elderly man chose the men to escort us, gave me a good gray-white horse and we started on the long and perilous journey. We passed many ruins of ancient cities, including a few rather large ones. We breakfasted in the morning around 5 or 6 o'clock. We had three tents; breakfast was served in the third. The table was covered with clean cloths. Fried eggs, tortillas and mutton with salt, pepper and mustard made up our meagre breakfast. We ate a lot of peppers. At 12 noon, we took our second meal, which consisted of tortillas rolled in paper, roast mutton, oranges and jam. We had English beer with us. The kitchen, along with all the kitchenware, the tents, the iron beds, the rugs, pillows, mattresses, duvets, etc., everything we had to carry along ourselves, because, except for dirt, there was nothing in the Bedouin tents. They always walk barefoot and have no clothes, just some dirty rags partially covering their nudity. They wear a kind of cap on their heads to protect against the hot rays of the sun. On the first night, we had just set up our tents and the Bedouins had just sat down around the fire they had kindled with bushes, when suddenly one of them leapt into the air yelling because he had been bitten by snakes. But fortunately, that turned out not to be the case. The snakes had wanted to approach the fire and they had crept under the man's bum. They killed them and thereafter we were often startled by the Bedouins' cries during the night. It is a wonder we were not bitten during the journey, [35] as poisonous snakes and scorpions often entered our tents. It was even more dangerous because we always had to go outside the tent at night to urinate, because there were no bedpans. Finally, we arrived at the Rocks of Petra. It took us the whole day to climb to the top, for it was a terrible path, often only a foot wide, slippery and tortuous. It was no more than a narrow path that led us along the cliff's precipices, constantly with the threat of death, because at each step we saw the deep abyss below us and we had to close our eyes not to be overwhelmed by fear of heights.

Madrid Wednesday 7 September

Today I went to the Alameda,[201] property of the Duke of Osuna.[202] I took a carriage because the Alameda is 1½ miles long. The coachman stopped at the end of the park and I still had a quarter of an hour to go on foot. The doorman showed me the inside of the palace. The rooms are small but well-furnished and all the walls are covered with beautiful frescoes or with silk carpets and are hung with

201 An avenue lined with poplars.

202 Mariano Francisco de Borja José Justo Téllez-Girón y Beaufort-Spontin (1814-1882) XII Duque de Osuna.

paintings and lithographs. On the brick cloister in the dining room is a parquet that is a successful imitation of the famous mosaic in the museum of Naples, which depicts the battle between Dario and Alexander the Great. There is another room, which houses a billiard table; another room has a display in wood of a bullfight; a matador holds his cape in front of the raging bull and waits carefully for the right time to place the sword in its neck; another matador has already stabbed the bull's neck with his sword; another man comes with his team of mules to lay a noose about the neck of a bull killed.

[36] There is a wonderful chamber belonging to the Duke with mattresses with white linen, duvets and pillows, with a porcelain sink, toothbrushes, etc. All the rooms have fireplaces with tongs. However, it all seemed a little dilapidated. Most of the rooms are covered with sheets and cloths. Afterwards, the gardener accompanied me to the garden, which is beautiful on this arid soil. But despite all the care, the lawns have been scorched and the leaves on the trees withered by the hot sun. There is a beautiful beehive, made so that, by opening a little door, you can see through the glass how the bees work, how they make their wax and deposit the honey in the cells. If I had been pregnant I would have miscarried, because the gardener had me enter a cottage where a hermit made of wood sat reading his Bible. He led me to another house where an old codger, his wife and grandchild sat around a circular table; before them were bread, roast chickens and two glasses that seemed half full with wine, but it was all mock and the red stuff just looked like wine. Upstairs in the same house were rooms. There was another cottage with a small collection of brass cannons and nearby there was a small fort surrounded by a moat. There was an orangery where they put the exotic and tropical plants in the winter. The oranges for example, used to great warmth, do not tolerate the cold. There was a dance pavilion and a few ponds where swans and ducks swam. [37] In front of the big pond was a pavilion where gentlemen prepared their rods and lines for fishing. There was a fountain laid with small stones and pebbles so as you don't get wet if you're on a bench under the fountain, whereas everyone else at a distance does get wet. In a wall niche stood a small statue that had been damaged by a couple of barbaric visitors. The only heir to the Duke, who is not married, is his cousin Prince Angrona Marquis of Javal V. The Duke is called Piedro Alcántara Tellez Giron.[203] In the garden, he has erected a monument to his grandmother Maria Josepha Pimentini, 205 Duchess of Osuna[204]. The administrator of his farms is the Marquis Alcanizes in Alcala Street. The Duchy consists of 1422 villages and 300 cities.

Madrid 28/9 September 1859

Yesterday morning, I packed my clothes, paid my bill and, after handing my luggage to my servants to be kept until my return, I left by train with my teacher. It was the festival of the Blessed Virgin and there were a lot of people about who wanted to take advantage of this day of rest and enjoy the countryside. We had good company in our section of the coach, among them Miss Maldonado, Rafael G. Maldonado's sister, who lives in Madrid, Valverde 46 3. The sister is a student

203 Pedro de Alcántara Téllez-Girón (1776-1851) IX Duque de Osuna.

204 María Josefa de la Soledad Alfonso-Oimentel y Télles-Gitón (1750-1834).

Figure 14: Alcantara Bridge at Toledo 1889

of the College of Noble Maidens in Toledo; the brother handed me his card and said that if I presented this at the college, and mentioned the name of this young lady, they would show me the college.

The streets of Toledo, city of about 15,000 inhabitants, have a medieval layout, for they are narrow and winding, while the façades and interior arrangement of the houses are irregular. The cathedral, which the Spanish believe to be one of the world's most famous because of its paintings, statues and ornaments, should be visited before one visits Italy, because all the most important cathedrals [38] of that country are superior in every way. The altar and the choir are surrounded by beautiful stone carvings and enclose wooden adornments of fine quality. The balustrades are wrought iron. In a chapel the Blessed Virgin with her Son in her arms, a sculpture of pure silver, stands on a silver cloth-covered altar. The windows are decorated with beautiful paintings; the arched vaults are plentiful, but the whole church could be placed on the floor of the Cathedral of St Peter in Rome and its grandeur would seem small in the gigantic space, for it would fit four or five times into the Cathedral of St Peter.

For lunch we had roast meat, boiled eel and fruit. Afterwards we went to the Alcázar, the palace built for Charles V. Standing atop a hill, this palace seems quite grand, but inside it is mostly in ruins and the stucco ceiling and arches have collapsed. There is a spiral staircase of great quality, because under the one spiral staircase is another that makes the same curve. The guard, a soldier, took us to the downstairs rooms where the stables are located, which could hold 25,000 horses. There one also finds the hearths and kitchens. And then we went to the insane asylum. There are about 40 women and the same number of men. Among the women there is a Marquise who has lost her mind by reading too many novels. She seems to be about 35 years old and she answered my questions in a confused way. She has her own maid. The women are separated from the men. They sleep on the floor on mattresses and have white sheets and pillows. The rooms are large, but there is no furniture. In each room is a fireplace, secured by curved iron guardrails, so that the insane will not burn themselves. Then we went to the girls' college, [39] where we were received warmly because Miss Maldonado had already spoken of us. The head mistress showed me all about the extensive building that had been founded by a cardinal during the reign of Philip II [1527 - 1598]. The Cardinal's ashes rest in the church. The building looks much like an exchange, because it is open in the center and only protected from the sun and rain by canvas. Above and below are large corridors to stroll in the afternoon. We went back to Madrid on the 6:15 train. In the coach was Mrs Alfilar Alfaro, of Carretas Street 15, 2nd floor, with her young daughter of 16, who is rather lively but less beautiful. I flirted with them and flattered them such that they both licked their lips with joy. This type of girl does not ask a man if he has lumps in front or behind; or if he is hunchbacked or bent, or if he is old or young, handsome or ugly. The only thing they want is to get married as soon as possible, because, unfortunately, there are many more women than men, and especially far more women than men can support. The mother gave me her home address and I had to promise to come and visit after I returned from the farm [Granja de San Ildefonso].

Madrid 9 September, 1859

St Petersburg
G. Salapaka & Son

My dear Sirs!

Confirming my latest letter from Paris, dated 12/24 last, I have the pleasure to say that I received yours from 14/26 last, in which I noted that you have recovered your letter of credit worth 160,000. I also take note of the calls to bid for 500 shares at the rate of 111. If the price of the shares in the Dunaburgo – Riga[205] railway line rises, I will instruct you about the sale. Two hundred shares cost me 3¼ and 1000 shares 4¼ premium. If it is announced that payments can be made on the shares of the said railway line, you will immediately hand over the payments on my shares, and if you haven't [40] any money of my own, then you may deduct a portion from the letters of credit to meet the necessary funding.

I thankfully note that you have deposited 7979.30 to Mr U. Zellweger & Co. for which I had reimbursed you 2174.20. After their arrival, please send the 30 tons of rubber to A.F. Matveief in Moscow. I reimburse you 19.99 paid to Mr G. Chamot.

I am very happy with your interesting information. After my return to Paris, it would be my great pleasure to go to Sweden to discuss the properties that you have suggested with your son and brother. Please send me Mr Sterky's address

205 This place was called Dünaburg in German. Today it is called Daugavpils. The railway line between Daugavpils and Riga was opened on September 12, 1861. It is 218 km long.

in Sweden. If an interest rate of 4 or 5% is likely, then I will buy in with great pleasure, but if the profit is not certain, then I will not buy in, because there is nothing which vexes me more than working for nothing.

San Ildefonso 29/10 September

We left Madrid last night at 11 pm and we arrived here this morning at 7 am. We had good seats in the berline[206] for which we had to pay $8 per person. Instead of finding – as I hoped – a regal abode, I encountered a miserable one. The royal residence is more like the house of the village chieftain than a royal summer palace. Everything on the ground floor is dirty and seems to be in a state of disrepair. It is said that the upstairs rooms are decorated elegantly, but I didn't manage to see them. We couldn't get permission to see the gardens, [41] which were closed today because of the lunch the queen was hosting and to which more than 300 people were invited. That is why we went by bus, for 1/2 $ p.p., to Segovia, one of the oldest cities in Spain. There the famous Roman aqueduct can be found, more than 150 feet high, consisting of large arches built on top of each other. The beauty of it is that the stones are joined without mortar or cement. It is unbelievable that arches made thus could carry such a heavy burden for so many centuries. In addition to the aqueduct, the cathedral deserves admiration for its splendid arches. The balustrades that encircle the sacristy, the choir and the main altar, are of wrought iron and beautifully ornamented. In the choir are many medieval religious manuscripts, written with a quill. The marble pulpit is beautifully sculpted. After that we went to the Elkasar [Alcázar], which is a small fortress palace with many cannons. The castle is very old but recently restored. It contains beautiful halls with gilded and multicolored ceilings in the manner of the Moors. In the halls are all sorts of architectural features. From the balconies, one has a beautiful view of the valley. Then we drank cocoa in a café where everything was filthy. We took the same bus and had outdoor seats where we were exposed to the full force of the burning sun. Once we got back we had a new problem for we didn't know how to proceed, since there is neither a bus nor a carriage going to the Escorial, and to travel the 7 miles to get there by coach they charged $30, which I stubbornly refused to pay. I arranged a good saddled horse for $4 with a woman, and her husband would accompany me on foot and would take my bag. [42] Or, for $6 more, they would give me two horses if my companion would agree to come with me. It is indescribable how annoying I find this journey. It seems that we have been sold and we'll just have to allow everyone to rob us at will. On the bus back from Segovia, a couple of officers sat behind us.

El Escorial Sunday 30/11 December, 1859

Yesterday evening I went to bed at 6 because I was very tired, only to finally awake at 6 this morning. At 7.15 we went to the royal gardens, which rival in splendor even the famous gardens of Versailles. The cedars around the flowerbeds were pruned in the shape of a square or a ball. Amid the flowers the queen of

206 A carriage in which the front and rear compartments are separated by a glass partition.

independence sat on a rock, surrounded by many other gods and goddesses with bouquets in their hands and pipes in their mouths from which water flowed. There were many such fountains in the flowerbeds and on the crossings of two paths and nothing was more beautiful than walking among so much splendor, as the fountains flowed. As I had separated myself from my companion, I left the path and, not without difficulty, reached the large pool by which yesterday the queen had held the grand lunch. There were still flower pots and vases with garlands. The gardeners there explained to me that there were more than 600 attendees, too many for all to be seated at the table. That's why everyone except the Queen lunched standing. The champagne flowed like water. Because it is surrounded by mountains and full of holm oak and ash, this place is the coolest of all the royal abodes and the favorite of the court. At 9.30 or 10.00, we left in a carriage which resembled [43] the kind of medieval coaches often seen in old paintings. We went continually uphill to the pass for four hours and then descended rapidly, and we arrived at the Escorial at 5.30 in the evening. We had contracted the coachman for $14 if it remained only us, $13 if he rode with someone else on the bench. He was not satisfied with the $13 we handed him and asked for $14, saying that he had only taken someone along on the ride who belonged there. He did not want to accept the payment and went to complain to the mayor, who sent a servant with the order for us to report immediately. The judge decreed that we had to pay half of the difference and half of the subpoena, which we did without complaint.

Figure 15: Alcázar de Segovia 1886 – by Juan Laurent Minier (1816-1886)

El Escorial 31/12 September 1859

This morning at 8, we left with a professional guide who had been blind for 43 years and who nevertheless took us everywhere and provided ample descriptions. You might say that the blind man saw better than the two with sight. A beautiful building that attracts the curious from far and wide, it forms a large parallelogram, the façade being the palace, the two annexes the monastery, and the church forming the rear. This grand building harbors 10,032 windows and doors, 16 patios, 9 towers and 76 fountains. We visited the palace, whose rooms are characterized by the immeasurable abundance of hand-embroidered carpets and stunning paintings. The furniture and the frescoes on the arches and ceilings, the paneling of inlaid woodwork, everything gives the air of royal splendor. The four rooms that make up the cabinets and the queen's boudoir are especially known for their opulence; each lock is a masterpiece and each door worth a fortune. [44] Afterwards we went to see the entire church and its fixtures. Likewise, this building is a beautiful piece of architecture. For example, there is a hall whose dome is as flat as the court. The pulpits are made of white and yellow marble. There is an inexpressible number of paintings, all masterpieces. There are four organs. Above the main altar are bronze statues of the Apostles. However, I must admit that the bare walls and brown-grey columns are a bit disappointing. The most striking of what I saw is the pantheon of the kings, which consists of a circular room with granite in the walls, and all around niches with granite tombs that seem to be carved out of the walls themselves. When a king dies and his funeral takes place, the cover is removed and the body placed within. It is a special feeling to find oneself surrounded by the remains of so many kings for whom the world was too small and who now lie quietly in their cramped coffins. How many masses must be celebrated to bring peace to their turbulent and revolutionary souls. Since today is the anniversary of the death of Philip II, there is a catafalque in the church, covered with a black cloth with a crown on it.

Figure 16: Escorial

We returned to Madrid by carriage. We had seats in the rear and were shaken by the bumps in the road that the carriage hit. After arriving I went into a tobacco store to buy cigars and I laid my cane on the counter, from where it suddenly disappeared. I was very angry, because I was very fond of that cane.

Córdoba Friday 4/16 September 1859

I left Madrid on 2/14 with the noon carriage. I had a coach seat and could see out very well, although I had to suffer a lot of dust and the discomfort of rising and falling. [45] The carriage was placed on a train wagon. At 11 am we arrived in Tembleque, from where we continued the journey by coach. We left very quickly to stay ahead of the coach from Madrid, which had left the same time as us. The jealousy that feeds this rivalry is quite good. The horses were ready and were hitched in an instant. We passed through La Mancha, land without trees and very flat and sparsely populated. One can travel 2 or 3 miles without coming upon a village. There is no shortage of beautiful women in the few towns and cities there are. The novel about Don Quixote has bestowed an indescribable appeal on this country. During the night of Wednesday to Thursday, we crossed the Sierra Moreno and even though it was night, we could recognize many beautiful landscapes by the light of the moon. We passed by Madridejas, Valdepeñas, Baylen [Bailén], Andujar, etc. Although you can see well from the seats we had, they are dangerous for those who can't hold their body weight while hanging onto the leather straps. In addition, only a leather cushion covered the wooden bench, which is painful to your backside. We arrived in Cordoba yesterday at 9 o'clock; it is a fairly clean city but with very bad paving, so a coach can only ride at walking pace. The narrow streets protect against the hot rays of the sun. The indoor design of the houses is similar to those in Pompeo. In the homes of the upper middle class, no windows face the street. There is only a big door, decorated with large nails. Behind this is a pretty little garden with cypresses pruned into fanciful figures. The dwelling is supported by pillars surrounding the garden, thus forming large and wide corridors where it is always cool. In the homes of the rich, the patios [46] are shielded by tarpaulins. Within the rooms, there are windows on either side, so they can be kept cool by drawing the curtains. There is no shortage of bedbugs and fleas. Onward from the Sierra Morena where Andalucía begins, the soil is far more fertile and rye, onions, oats and wheat can be grown everywhere, but the land is especially suitable for olive trees, with which it is covered as far as the eye can see. The olives are harvested in January and although this year's harvest was not so good, generally speaking, there's no better way to make a profit. We stayed at the Rizzi inn, the only one in the city. Despite the many bedbugs and fleas, I slept well and this morning I took a bath in the Guadalquivir. After dinner, I went to the famous cathedral whose dome rests on more than 800 pillars that were originally from the ancient pagan temple of Juno which was situated on the same spot. With this mosque, the Arabs wanted to create one of such grandeur and beauty that it would surpass that of Mecca. In it are some chapels which have been preserved, including those where the Capialla[207] was. There are still many inscriptions in Arabic. The stones have been worn and are concave where the Arabs have wiped their feet. There are 19 altars and the main choir, whose wood carvings are of great quality. They depict the saints of Córdoba and the history of the Bible. There is a chandelier made of pure silver that weighs 26½ arrobas. In the columns, made from porphyry, are carvings that depict the Holy crucifixion and it is said that these are made by prisoners, scratched in with their nails. And that is not surprising, considering that a kiss that is softer than a nail eats away at the porphyry. One of the better inscriptions is protected by small wrought iron bars and next to it [47] is a bas-relief of the prisoner. The beauty of the markings is that the prisoners made them behind their backs, as they stood with their hands tied behind their backs, chained to the pillars. In the courtyard of the mosque are beautiful avenues lined by orange trees. The waiter had some sort of swelling on his instep; I asked him what ailed him and he told me that his varicose veins were protruding.

Seville Saturday 5/15 September 1859

Yesterday we left Córdoba at 6.15 by train. We went very slowly and experienced much ado, for example 3 quarters of an hour in Posada, making us not arrive in Seville before midnight. But on the way, the conversations we had with the women, especially in Posada, made up for much. In the same coach with us was master carpenter Juan Causino Nepomuzeno, of S. Francisco Square no. 7, who told me about a farm for sale at 1 mile from here, which, by his own account was a good estate, both in practicality and for enjoyment. He promised to send his son this morning, who would guide me. He came, but immediately pointed out major differences from what the father had told me. He promised to ask permission to view the farm tomorrow morning and we would go together on horseback. We went to the cathedral, which is one of the most beautiful I have ever seen. It has very high vaulted ceilings and 91 stained glass windows. The sacristy, the choir and the main altar are woodwork masterpieces. The crucifixion of Christ, also in wood, stands out. Everything is surrounded by bronze or wrought iron bars. In the Royal Chapel, the bodies of Fernando[208] and other kings are honored. An adjacent chapel is famous for its wealth in silver, of which the stiles, the altar and the large chandeliers are made. Among the many paintings there is one that the English [48] had promised to shroud with ounce gold, if the church had wished to sell it for this price. Many of the ornaments are gothic-style replicas. The church is 431 feet long, 139 high, 171 in the dome, and 315 wide. From there we went to the art museum, that had but two halls, of which only one had paintings by Murillo.[209] Next we saw the gardens of the Alcázar, full of fine paths lined with

207 Here Schliemann writes in Arabic the word 'Canela', which means cinnamon. He probably meant Capialla, from Capillia de Villacicosa. The Mosque & Cathedral of Córdoba also known as the Great Mosque of Córdoba and the Mezquita, whose ecclesiastical name is the Cathedral of Our Lady of the Assumption.

208 Fernando el Santo (1199-1252).

209 Bartolomé Esteban Murillo (1617-1682), who lived and worked in Seville. In the nineteenth century, he was the most famous painter in Europe.; his work was worth more than that of Rembrandt or Vermeer.

orange trees, myrtle, etc. Everywhere you look are fountains to water the plants; the gardener opens the valves with a push of his hand. Then we saw the Alcázar; its outer walls and the chambers within are both worthy of a journey around the world. The walls, the halls and the ceilings are covered with Moorish stucco bas-reliefs, interspersed with verses from the Quran, all painted in very vibrant colors intermingled with gilding. The courtyard is indescribably beautiful. There they let me see the room where the Duchess of Monpensier Orleans, sister to the queen[210], gave birth to a child in 1848. No doubt she had chosen this particular spot for childbirth to lend even more glory to this illustrious birth, probably in the hope she would give birth to a son who would ascend the throne of Spain. It was raining so I went home to write about the remarkable things that I had seen. Nothing is more beautiful than the small narrow streets, mostly paved with flagstones and very well tended, and the pleasant brick houses, whitewashed to reflect the heat of the burning sun. The dwellings usually have no more than two floors. On the ground floor are two windows, protected on the outside by green-painted grating and inside by curtains. In the center of the house is a large door made of wood painted brown or a green-painted gate that opens to a wide corridor leading to a courtyard of marble flagstone. In the middle of the courtyard is a fountain. The courtyard is overhung with awnings that can easily be made smaller or wider by means of a rope. [49] Around the courtyard are lovely arcades that serve as living spaces during the heat of the summer and are furnished with chairs, settees, canvases, bird cages, flower pots, exotic shrubs, lanterns – all this done in fine taste. The corridors are supported by pillars and at every step the guest stops in front of the balustrades or the open doors to admire the beautiful courts and enjoy the view. On the second floor are no windows, but glass doors which open onto small balconies.

Today, as I crossed the fruit market, I met a woman who was accompanied by a rather pretty girl of 14 who however seemed sad and melancholic, which triggered my interest. I approached them and asked the old woman to give me the little one to live with me in Paris. She answered that it was imperative that we knew each other a little better first. She gave me her address and in the evening, I went to her home. I had seen the two on the street and it had seemed to me that they would agree to the proposal and I had started to envision that there would be no obstacles. I was therefore very surprised when I heard that the daughter didn't want to go nor did the mother want to part with her for fear that the neighbors would criticize her if she abandoned her daughter, who was still so young. When I saw that all my urging would lead to nothing, I suggested I accompany them to the ball. They wished to go alone and I made no objection. I was softened when I saw my beloved sitting on the guest couch with an angry and sorrowful face. In vain I tried to make clear to her and her mother what a terrible future the poor child could expect if they refused my kind proposition, which had no other purpose than to secure the little one's happiness. But to no avail. In vain I labored to explain to her that she would be wooed and abandoned by a scoundrel and that she would regret for it the

---

210 Luise Fernanda de Bourbon (1832-1897).

Figure 17: A Gypsy Girl of Seville 1887 – by John Bagnold Burgess (1829-1897)

rest of her life. The ball was pathetic and I quickly left while following the old woman and her daughter. I treated them to sorbets, taking the opportunity to show them how good life in Paris would be. Fearing that my passion for the daughter [50] take root too deep, I wanted to take leave of them while the feelings of love were still tender and so said goodbye to them forever.

Seville Sunday 6/18 September

This morning I left with the broker in the direction of Bermuza to view Venen's farm. The man had boasted of his skill on horseback and I was stunned to see that he had difficulty trotting. He blamed the saddle, saying that he was not used to saddles other than Spanish. The ride was one of the most pleasant I have ever made in my life. The view of the mountains above the city is stunning. Also, I liked the farm. There are more than 4000 olive trees and two large vineyards. However, neither has yielded a good harvest this year; the olive trees produced scarcely a third of the olives they yielded two years ago. Last year they produced less than 35 arroba of oil, but generally you can count on an annual income of at least $1000 from the olive trees and vines combined, which is not bad, but it would be better if it were more. If I buy this farm, I do so mainly to make love to the beautiful Andalusian girls, but I'm afraid this pleasure would quickly tire, because I'm surrounded by imbeciles and there's no hunting or fishing to be had here. The ward's house is worthless, for it has but three small rooms and one would certainly have to add a level. All the machines used in the manufacture of the oil and wine are in good condition and the walls seem to defy the centuries. The supervisor seems to be an honest man and I wish no better. Except for a few orange trees, [51] there is no garden. We then went to another farm near the city which comprised approximately 800 orange trees, pears, etc. and sown among them corn, vegetables, rye, etc. There are two houses two storeys high and in good condition. I asked the real estate agent to inquire about the price of this farm. On our return to the city, we passed the grand gardens of the beautiful Palace of the Infante.[211] I then offered my companions a good lunch and, without changing clothes, went to the bullfight, where I had a bad seat and almost fainted from the heat. It seemed that the bulls here are wilder and that both the toreros and the picadors worked with more skill and speed than in

211 Infante: rank and title given to the sons and daughters (infantas) of the king, sometimes with the exception of the heir apparent to the throne.

Madrid. The excitement was so great that several people around me began fighting amongst themselves and had to be escorted out by the police. I left before the end and after dinner went to the San Fernando theater where El Cura was performed, a ballad in verse.[212] The seat in the courtyard cost but 3. The performance was very pleasing, as was the theatre, which was beautifully adorned. All seats in the courtyard were leather armchairs.

Seville Monday 7/19 September

Tired of the many entertaining things I did yesterday, I woke up late today and after breakfast I went to the cathedral where a redeemed monk pointed out all the important things, including the tomb of Cristobal [Christopher] Columbus' son.[213] An inscription on the stone slab declares he had dedicated his whole life to the arts. Having combined all the then known sciences into a work of four volumes, he bequeathed that together with his library to the city of Seville. He died at the age of 50. There are also stories of the two ships of Cr. Columbus, of which there are drawings. [52] They are small two-masted ships, with a combined crew of 90 hands. They were built to be rowed when the sea is calm, as shown by the many seamen holding oars. All Columbus's arms as admiral and viceroy can be found engraved here in the stone. After that I saw the cathedral's riches, consisting of silver pieces and priests' embroidered vestments, and then I climbed the tower. Instead of stairs there is a wide and gradually spiralling path which you could ascend on horseback if the door at the entrance was not so low. On the tower's outer circular balcony, where the bells hang, one has a magnificent view of Seville with its thousands of dwellings, whitewashed to reflect the heat of the sun. From there I went to the tobacco factory and a concierge showed me all the ground floor rooms. In one is a forge, in another a factory in which snuff is packed in tins. There is a room in which bales of tobacco are stored, in another they cure the tobacco with sugar and other ingredients and in still another it is ground by mule-drawn machines. Next we went to the second floor where gigantic rooms are full of tables at which women work. One picks the leaves from the stems, others roll the cigars, others make sealed paper bags for the cigars from a mold. At each table, only one task is done and thus it's done very fast and easy. I can't deny that seeing so many women was the most interesting spectacle I've enjoyed in my life. They are mostly beyond hope, but there are some very pretty ones amongst them. Judging by the hollow eyes and the look of venereal disease, I noticed that almost all of them were sick, which is normal because they cannot clothe themselves from the 3rs they earn per day and they must take to the streets to supplement their earnings. Despite their miserable lives and their still more miserable prospects for the future, [53] and despite their illnesses, they all seemed very cheerful and they

---

212 Pérez Escrich, Enrique (1823-1897) 1859 El Cura de Aldea: drama and tres actos and verso. Madrid.

213 Ferdinand Columbus (1488-1539) He was the second, illegitimate son of Columbus. As a young boy he traveled to America with his father; later he spent a large portion of the money he inherited from his father on a library of 15,000 books. He wanted his library to stay intact after his death, but that was only partially successful; 7000 titles came into the possession of the Cathedral of Seville, where they now form the Bibliotheca Columbina.

burst out laughing when they saw that I was eyeing one up and felt compassion for the other. Several times I asked the prettiest if they wanted to come to Paris with me, and all of them were ready and willing. The forewoman (who accompanied me to prevent unrest and scenes on the girls' part) coughed a lot, has a sunken chest and is incurable, for she is in constant pain and will soon die from consumption. Then I went to the market, where a large feast was held yesterday in honor of the procession, which arrived by rail from Jerez for the first time. The tables were still there. The market is built like all other buildings, with a courtyard in the center and surrounded by arcades supported by pillars. The whole second floor is full of rooms containing the archives of the Spanish provinces in America, from 1500 to now, and all reports and information given by the great explorers and generals, all the treaties made, etc. Each piece of paper is stored separately and everything is packed in parcels labeled with what it is about and in what period it belongs. Leading from the second to the third floor is an admirable staircase, built of large stones that were laid without the least support and without a stairway arch to support its enormous weight. The only thing I think could be giving it support is the fact that the bottom piece of the stone is cut to a slanted surface, like a windmill's blades. Next I went to the bank where I saw on a notice that the discount for bills of exchange was merely 6%.

Cádiz 8/20 September

At last I reached this beautiful city, which I've never thought of but with joy, because all who have been here have praised it to high heaven. [54] This morning at 8 I left Seville. The view of Seville with its beautiful Giralda[214] was very nice but when we finally departed on the Guadalquivir I felt seasick all the way to Córdoba. On both sides of the river were beautiful farmsteads. In Seville, the river is no wider than the Seine in Paris, but it gradually widens towards the mouth. About halfway between Seville and the sea, the ground becomes flatter and is unprotected from the river's floods. There is nothing but marshes and meadows as far as the eye can see, and untold flocks of birds, mainly ducks. On the steamship, we had good and cheerful company and because of the nonstop conversation, time passed quickly. In San Lucar most of the passengers disembarked; because of the ship's movements and seasickness they preferred to continue the journey by land, even if that meant a big detour. With its countless ships, all pavilions in the world and its splendid buildings, Cádiz presents a very nice view when arriving by ship, especially since all the space that nature has allotted the city is taken up by buildings that seem to be built on the water. I stayed at the Blanco Inn, which is but third class compared to the Paris Inn in Seville, though I have a view of the sea from my window. From 6 to 9 o'clock I took a walk, looking for romance. I spent 1½ hours on the beautiful boulevard where all of Cádiz was browsing to the sound of music. The young ladies seemed to strive to display their fair faces, large eyes, which shone like many suns, and the simplicity of their Spanish dress, with their long veils, to heighten the effect of their beauty. In the presence of so much natural

214 The bell tower of Seville Cathedral, originally built as a minaret.

Figure 18: Cadiz ca. 1850

[55] extravagance, I couldn't imagine befriending one of the unfortunate women I encountered at each step and who were ready to haggle about the price of their love. Twice I approached respectable ladies, but they did not answer and later I no longer dared to try to start up a conversation.

Cádiz Wednesday 21 September

I was so tired from the many roundabout ways I had taken yesterday, that today I didn't wake until 7.30 and got up at 8. I went to the sea spa, where I enjoyed my bath very much. The sea water contains so much salt that swimming takes less effort. Afterwards I dried myself off with a towel, ate a breakfast of fried eggs and tomatoes, grilled fish and cocoa and then went on my way to tour the city. I took a walk along fortifications about the city, crossed the boulevard where I had heard music yesterday and had seen the lovely maidens; the view of the sea, which evokes so many memories, always fills me with joy. Everywhere, cannons are positioned to defend the city, which was once plundered by the French and later futilely besieged several times. There where openings had been made to repair the fortifications or to plant trees, I saw that the ground was full of rubble from tiles, pottery, stones, mortar etc., which is why it's difficult to have gardens here. Further away are two castles, one of which is built upon the rock; they say that the foundations were laid by the Phoenicians. Next I criss-crossed the city in all directions and everywhere I went I met the most beautiful women, dressed in black serge with a large scarf of black silk that fell to their backside and a veil wrapped about the head. Their black or blonde hair, that would have reached the ground if it had been loose, was bound behind in thick braids. They have very small feet. Even though the streets are very narrow, they have flagstone sidewalks on both sides. The air here is refreshed by the sea breeze. [56] The houses are very beautiful, some with flat roofs and some with sloping tiled roofs. Although Cádiz seems to be a paradise, I wouldn't be able to live here without some solid endeavors, which I wouldn't be able to pursue without having to pay much for my schooling. On the walls that surround the citadel, hundreds of men are always present, fishing with rods or lines or nets.

Cádiz 22 September (Thursday)

This morning I left at 8.10 in the direction of Jerez. Since they hadn't finished laying rails to the dock, we had to cross the bay to the railway's pier in a steamboat. The pier is located at the spot where, in 1810, the French positioned their artillery and bombarded Cádiz. It's a long distance and it seems impossible that the bombs could have reached the city. We traversed level ground in which many drainage ditches had been dug. Shortly after we saw, on both sides, nothing but vineyards which produce the delicious sherry. The soil was sandy but nevertheless the vines appeared to grow well. The city of Jerez has absolutely nothing interesting to offer except wine bodegas. I saw the French consul Domecq's[215] wineries, which are very large buildings filled with barrels of wine. The servant ladled a little wine from various barrels and let me taste a wine that was 60 years old and as thick as syrup. I saw how they make the vessels as well, how they mount the cast-iron rings, how they burn the inside, etc. I visited the casino, too, and went back to Cádiz at 2 o'clock. That evening there was music on the plaza near the boulevard and there one saw the flower of feminine beauty. I approached one and, with effort, managed to start a conversation that became more intimate once I started talking about women's dress. [57] I left once I realized that it was impossible to gain special favors, but not without jotting down the address of the girl named Eliza, Calle Larga No. 50.

Algeciras 23 September (Friday)

We embarked this morning at 6.30 on the steamer Bernardo, direction Gibraltar. With sorrow, I took leave of Cádiz and I climbed up the mast to look at the city for as long I possibly could. A voice told me that I would never see it again. The steamship, which had but 25 hp, went very slowly. Cádiz disappeared and soon we saw the mountainous coast of Africa. The wind, which had turned and blew with force through the narrows, delayed our trip even more. Fortunately, the current helped us, otherwise we wouldn't have made any headway. We rested for a while in Tarifa and arrived in Algeciras that evening at 5.30, in whose port lay several Spanish steamers that had brought troops and ammunition for the war in Morocco.

Gibraltar Saturday 24 September

I arrived here this morning at 9 o'clock and, after breakfast, I set forth to see the massive fortifications of this eagle's nest. Unfortunately, a permit was needed, which the governor of the city only extends at the request of the consul. As he

---

215 The company was founded in 1822 by Pedro Domecq Lembeye (1789-1839); following his death it was run by his cousin Juan Pedro Domecq Lembeye (1796-1869).

was not at home, we were not going to get it before 11 o'clock. Next I went with my Jewish guide to visit the underground passageways, one above the other at an angle, carved in the rock that overlooks the isthmus that connects Gibraltar with the mainland. The entrances of the corridors where the guns are positioned seem very large; they need to be large to be able to aim large guns. From below, because they are positioned at a height of 500 to 700 feet, they appear no larger than pigeonholes. Hence it is difficult, if not impossible, to reach [fire at] them. Lastly, one proceeds down into the underground tunnels, a descent of about 100 feet, and comes to a large circular room named the George Hall, [58] which houses about 20 large-caliber guns that dominate the sea near the strait. Next we climbed to the top of the rock, where a huge battery of large-caliber cannons stood. The impact of these artillery shells must be huge when launched from this height of 1460 feet. In the event of an attack, I would rather be on the rock than in the ships, because the force of the projectiles, increased by the almost vertical fall, should be such that the shells would penetrate the strongest ironclad ships. Beneath are also, both outside on the rock and in the village, several very strong batteries intended to defend the narrows and so the fortress is impregnable from this side, barring betrayal. After that we climbed an adjacent rock whose top was approximately 1250 above sea level. Although there is a road that gradually twists and turns upwards, it is very difficult and tiring to climb and we were sweating as if we'd spent the last hour in a Turkish bath. The view is beautiful and embraces not only the mountains of Spain, but also the mountainous coast of Morocco and the vast expanse of the sea. There is a small division of three or four large-caliber guns and a house where three guards live, who announce the arrival and departure of the ships by hoisting balls and flags. They continually peer in all directions through strong binoculars and no sooner have they spotted a ship in the distance than they have warned of its presence by hoisting a ball. While I was there, eight large English steam warships approached the port and cruised in; they all had 76 guns and the imperial one even had 141. The imperial one was saluted by the artillery with 17 cannon bursts and it answered with the same number of salutes. These steamships came from Malta. [59] The guards were very amicable, especially after I drank a bottle of beer for 15 rs. One of them lived there with his family and two children, one of which was born there. It was only 11 months old but was already walking and talking with no trouble at all, which proves that the climate there is particularly healthy. In front of the house is a fountain. If you look down from the cliff on the side of the sea, it looks like the rock is accessible, but that's just an illusion because there is 600 feet of sand, then 400 ft of sheer cliff and finally 250 ft of rock to climb. Then we went to the third cliff, the steepest of the three. At the top, there is no building but an old tower built by the Moors. On this cliff at the seafront and opposite Morocco is formidable artillery with 74 large-caliber guns. We went down and walked along the cliffs below, but you can't go farther than the batteries called the 4th Europa, positioned at the edge of the cliff where the Moorish tower stands. There stand the 3rd, 2nd and 1st Europa and many batteries with other names. Between the 2nd and 3rd Europa, the governor's wooden house is protected from the winds from the North and West by the cliffs. Except for a few plants whose leaves are used to make rope, there is no vegetation on this rocky ground, which has very little topsoil. We went back and again walked along the batteries on the seafront, after which we climbed the big cliff that towers nearly 400 feet over the batteries below, about 500 feet above sea level. Around this large plain above are large-caliber guns. There is also a large barracks surrounded by walls, with openings for the riflemen and some 35 gun embrasures. We descended a spiral staircase and continued our visit to the batteries that lay on the side of the bay. Although there are many well-built guns angled to fire ammunition, the batteries couldn't provide very strong [60] defense because of the large coastal area, and a still larger area where there are no bulwarks at all. It seems to me that an army could be brought there to disembark and take Gibraltar, for all the artillery in the passageways and all those who defended other areas would be of no use if Gibraltar were attacked from this side. The English have understood the necessity of a line of defense there and are already building more bastions, but the work is not yet done. In the center of the village is the governor's park; everyone has access to this park which in the evening is a popular walking route for the residents and, as one told me, there is always music. The city is well built and has some charming buildings. There are some straight cobbled streets with narrow sidewalks on either side, but, as the city is built on the slope of a hill, it's normal that most of the streets are steep and winding and many of those above are surfaced with sand instead of stone. Since Gibraltar is a free port, I expected to get almost anything here for practically nothing, but I was sorely mistaken as there was little difference in the price of cigars with the price in Spain, and shoes and clothing are even more expensive. That evening I went to the theatre, built of wood and very small. They performed a sung zarzuela [an operetta]. A passenger who accompanied me, and who is to leave for Malta today, had bought second-stage tickets, where only whores sat. We talked a lot with these unfortunates, but at 10 pm I left because I was tired. A ticket cost 5r.

At sea between Malaga and Gibraltar Sunday 13/25 September

We left this morning at 7 with the steamer San Bernardo in the direction of Malaga and when we rounded Gibraltar [61] I saw its beautiful bulwarks again.

Málaga Monday 14/26 September

In Spain, they have the bad habit of breastfeeding children up to the age of two and a half or even three years old. In my opinion that is very bad, but since the great heat often causes inflammations in the brain during the time the children are teething, it may be that they breastfeed as long as they do to ease the growth of the teeth. There's nothing remarkable here, only the lavish amount of fine fruit that fill the streets and town squares; the most beautiful grapes are sold for 3 Cop[eken] per pound. Yesterday after my arrival, I hurried to the post office, where I discovered my name on the list of letters. I asked for my letter and the clerk asked for identification, which I quickly presented. It was a letter from my agent in Moscow, which I answered today. I went to the theater again in the evening and despite everything, I bought a ticket for in the second class section, where I had to warm myself between two whores, one of whom seemed to be the madam

Figure 19: Gibraltar Clearing Storm 1860 – by Samuel Colman (1832-1920)

and the other the subordinate. Their company disgusts me. After I returned from the theater, I went to bed, for I was tired because for several nights I hadn't slept well. But no sooner had sleep overpowered me than I woke up with terrible pain and an itching all over my body. I lit the candle, examined the bedlinen and saw with horror that there were bedbugs of every size and shape everywhere. Because I hadn't brought insect powder with me, I slid two chairs together and lay on them with my sleeping bag as a pillow. But since there were rims on the chairs, this improvised bed turned out to be too hard, so I laid myself to rest on the two tables I'd slid together. But, as one can understand, I slept very poorly. This morning I threatened to go to another inn if they didn't give me another room without bedbugs. They gave me another, but I don't know whether it's free from bedbugs. [62] The rooms are like those in Cairo, only much filthier. A mosquito string (flypaper) hangs above the bed, which is necessary as there are many of these insects that bite and are a real nuisance. This morning I went to the seaside spa, which is very good. They gave me a bath towel to dry off with. Then I wrote a couple of letters, ate and went to the theater.

Granada Thursday 29 September (the day of San Miguel) 1859

I left Malaga at midnight of the 27/28th by stagecoach, which is a miserable cart with two wheels, with up front, the coachman, the postilion and a passenger, and four passengers behind in seats where barely two could fit. I was one of those latter unfortunates and I can't describe how much I suffered from the heat, the dust and the constriction because there was no place for your feet. Lastly, the company was also poorly chosen and I had to restrain myself in order not to offend someone's stupid pride. The first three hours out of Malaga were spent climbing a fairly steep side of a mountain, during which we had an almost unbroken view of the lights of Malaga. After that we alternately climbed and descended until we finally reached the beautiful plain of Granada. We passed the village of Santa Fé, where Fernando the Catholic[216] defeated the last Moorish king and took Granada, and half an hour later we arrived in Granada. I had planned on alighting at the Victoria Inn, but the owner of the Minerva Inn sat in the stagecoach, along with a tourist guide for foreigners, and upon my inquiry they told me that there was a Frenchman at the inn who, judging from their description, had to be Melchior Mocquai, with whom I had traveled from Madrid to Seville. [63] So, I preferred to stay at the Minerva and I'm quite content here. After dinner, I went to a whorehouse where I slid an enormous condom over my penis and fucked a rather pretty young girl, which I only did to be able to say that I had enjoyed a Castilian lass. After which I washed and dried myself off, returned to the inn and went to bed. This morning, after a cup of cocoa, I went to the Alhambra which is on a hill a little way outside the city. The entrance to the park is majestic, adorned with beautiful ash and other trees. While climbing the path, I passed the gate called the 'Gate of Judgment' [Puerta de la Justicia] on the left-hand side and entered a large open corridor, which led to the archives' little green. Nearby is a large plaza called Alcazaba and that of course was filled with age-old buildings.

Next I climbed to the top of the watch-tower, which had a spiral staircase and from where you had a panoramic view of the city and its surroundings full of farms, gardens and plantations, and of the Sierra Nevada covered with everlasting snow. I then descended and went to the Arabian palace, next to which is the Palace of Charles V, that was never finished. The outside suggests a luxurious interior, but if you climb the stairs you come across a giant round courtyard surrounded by arcades supported by 32 marble columns underneath and by an equal number of columns above. There is nothing to see downstairs because none of the rooms is finished. Before climbing the watch-tower, I had met a guide who I had hired to show me around for 5 fr. till 5 in the afternoon. I entered the Arabic palace from a 27-step staircase, all of which were 3 varas wide and all made of a single piece of jasper. I entered the Patio de los Arrayanes[217] or Court of the Myrtles, in which a fish pond is surrounded by myrtle bushes; on the left and the right are eight columns of white marble. Then I went to the Courtyard of [64] the Lions [Patio de los Leones], named so because of a fountain in its center with 12 Lions. The four sides of this court form a beautiful gallery supported by 32 small pillars of white marble. On both sides are pavilions with fountains in the middle. Each pavilion rests on 22 columns. On the right is the room that is called the Hall of the Abencerrages, in which 36 Abencerrages[218] were beheaded, so the story goes. In this hall is a fountain with a large brown stain which was supposedly caused by the blood of the beheaded. On the other side, there

216 Ferdinand II of Aragon (1452-1516).

217 Luma apiculata, Chilean myrtle.

218 An Arab family or tribe who played an important role in the 15th century in the kingdom of Granada. It is said that one of them fell in love with a lady of the royal family. He was caught when he climbed up to her window along the wall. The king had all Abencerrages in the Alhambra imprisoned and murdered.

Figure 20: The Alhambra 1815 – by James Cavanah Murphy (1760-1814)

are three rooms of which the ceilings are decorated with arabesque works featuring men and women. This seemed strange, especially since the Prophet Mohamed, to prevent paganism, strictly forbade the portrayal of human figures. In one of the halls, the Hall of Justice [Sala de la Justicia], a castle is painted from which maidens look out. On one side a knight spears another with his lance, on the other side a knight kills a wild beast holding an abducted damsel. In the other hall is a ceiling that also displays a castle with maidens. On the flagstone is a large porcelain vase. In the third hall the ceiling portrays a hunting party. On the opposite side is the Hall of the two Sisters [Sala de las dos Hermanas], named so because of two equally beautiful marble slabs; on both sides are small rooms; in the walls of the entrance are niches in which the footwear was placed. Then we went to the Hall of the Andaraje, adjacent to the Garden of Andaraje[219], that has beautiful fountains and orange trees.

There is a courtyard with wrought-iron grilles; the story goes that this was the jail where the queen was imprisoned. Next we went up to the Queen's boudoir [Peinador de la Reina] that was restored for Fernando and Isabel and has many paintings that have been destroyed because visitors have scratched their names in them [65]. In one corner is a pieçe of cast iron with several holes, which supposedly served to spread the perfumes and scents from the downstairs rooms. One has a magnificent view from the balconies of these halls. Then we went to the king's bath, which is a bathtub of white marble; in the same hall is the queen's bath; followed by the servants' chamber and then that of the royal children's bath. Then we entered the chamber where the kings rested after their bath. There are beds of marble, which in olden days were surely covered with beautiful rugs. In the middle of the hall is a fountain and upstairs are several arcades where music was played. As in almost all the rooms, the walls were decorated with Arabic inscriptions and the domed ceiling is wooden and very nicely painted. We went through the Hall of the Nymphs and came to the vestibule of the mosque and finally in the mosque itself, rebuilt into a chapel for Fernando and Isabel. Next we went downstairs to the Chamber of the Ambassadors and the Vestibule of

219 Most likely referring to the Hall of the Abencerrajes and The Garden of Lindaraja (or Daraxa).

Figure 21: Patio del Agua 1815 – by James Cavanah Murphy (1760-1814)

the Ambassadors [Façade of Comares]. All the walls are decorated with beautiful arabesques of painted or gilded plaster with a fine wooden ceiling. It is said that Isabel the Catholic[220] received Cristobal Columbus in this Chamber after his first trip to America. In the Chamber of the Ambassadors are nine niches with grilles. The entire building is in a state of ruin, the walls and surfaces are full of cracks; however, little by little, they are starting to restore it. It is the estate of the kings. The roofs are shared and are tiled. Then we went to the Cartuja [Monasterio de la Cartuja] which is located outside the walls and is a treasure chest of masterpieces. Inside is a gallery with paintings by a Carthusian monk, depicting the petitions presented by English Carthusians who did not want to submit to Henry VIII and didn't wish to acknowledge him as Head of the Church. In other chambers, painted on the walls are marble columns and a cross that seem real; only [66] when you look closely can you see that it is an illusion. In the church is an abundance of beautiful red marble, whose veins often depict people; there is also grey marble. The lecterns, cups with holy water, altars, everything is made of fine marble; the walls are covered with arabesques, the ceilings with beautiful frescoes. The doors are made of ebony and covered with shells, ivory and mother-of-pearl of rare quality. There are numerous ebony cabinets, richly decorated with mother-of-pearl, ivory and shells from the XVI century. We returned to the inn. My guide, a man without any education but consumed with pride, did not want to go on with me and asked 5 francs for his services. I had him given 2. After dinner, I went to the Generalife, the country house of the Moorish kings. I had to walk farther up on the magnificent lane that

220 Isabella I of Castile (1451-1504).

leads to the Alhambra. In the palace of the Generalife are two halls with portraits of the Spanish kings and some Arab inscriptions. Thanks to the profusion of water, the tiered gardens are kept beautiful. Afterwards I went to the Madrid banker Calderon's gardens, which were very well-kept. The trees were only a few years old, but already bearing delicious fruit. There's a pond with a tall arched bridge. In the middle of the pond is an imitation Moorish fortress, worn by time. On the pond swim two swans. There is a waterfall, a few sculptures and greenhouses for pineapples that are sold for $3 and $4; I quenched my thirst with pears and grapes and departed. [67] No sooner had I gone a few steps than I came across a gentleman who seemed well-to-do and with whom I started a conversation. He told me that he was going to the Church of San Miguel, which lay outside the wall, and I gladly accepted his suggestion to accompany me, because today is the day of San Miguel and there's a huge crowd of persons of both sexes. We had to climb a very high hill but we were generously rewarded by the sight of the beautiful young ladies hurrying uphill. At the top were a few little pavilions, in which refreshments were sold. I returned to my dwelling and was back in time for the meal.

Granada Saturday 1 October

As he had promised, the bookseller Carlos Gomez Moreno breakfasted with me yesterday morning at 10, and then accompanied me to the cathedral, a luxurious building from the XV century, decorated with fine marble, and the tombs of the Catholic monarchs Fernando and Isabel and of Philip I and his Queen Juana. Thereafter we went to the Alhambra, where I again passed through all the luxurious chambers and courts. They are restoring the palace now and it's about time too, for almost everywhere the colors are molted, there are cracks everywhere and the walls threaten to collapse at any time. The bookseller then wandered off to pursue his own matters, while I headed off to my compatriot Pablo Nottbeck, who is very popular here, both for his spending and for his amenity. He greeted me warmly and showed me the first miniatures he had made of the Alhambra, which he intended to exhibit at the New York exhibition. He then plans to present them to the Emperor of Russia, who will undoubtedly shower him with decorations, which he deserves for such a superb job which demands the great skill of both an architect and a painter. He led me to a book binder in whose house he also has a small workshop. Afterwards he was so kind to [68] accompany me to the Inn 'de los siete suelos', near the Alhambra, where he forced me to dine at his expense and didn't even allow me to pay my own share. During the conversation, I told him that in the 1½ years of my travels I had tried in vain to find a way to turn a profit with my capital. He replied that he knew of a good prospect, since he had a friend who was very ingenious and intelligent and who had bought a papermill across the river from the rapids (white waters) at Duda, one and a half miles from here. He said this wise man's name was Fernando Wilhelmi and that he was already pulling a monthly profit of 60/m rs out of his factory, to which he had introduced many new innovations. That he (Nottbeck) had unsuccessfully tried to persuade his brothers to offer this man a payment acceptable enough to entice him to St. Petersburg and to improve their papermill there. According to him) this is already yielding a 3000 ruble monthly profit, and that should easily double. Finally, the man told him that this is a land of sugar beets and that a sugar factory would break even in two years, because sugar beets grow here year round. I was pleased at such a prospect and expressed the desire to get to know this Wilhelmi. Nottbeck (Don Pablo) offered to escort me there by horseback. We departed, and after one and a half hours we reached the factory. The exterior of the building, with its tall chimney, gave me an impression of the factory's importance. Wilhelmi, although he didn't even speak his own language very well, seemed an honest German manufacturer who doesn't lack in knowledge of his art, but in my conversation with him I wasn't able to detect that superior intelligence that Nottbeck had sung the praises of. Then I discovered that he had ideas that would drive up the price of this company and that he constantly contradicted himself, because first he said that a factory would cost 400/m to process 5/m quintales of sugar beets and then that one could be built for this price that could process 200,000 quintales of sugar beets and [69] at 7% sugar, 14,000 quintales of sugar, the same amount of alcohol, a large amount of sugar syrup and fertilizer. He told me that for firewood there were plenty of thorn bushes to be had from the hills for little, and that the workers would gladly devote themselves to sugar beet cultivation if they were given seeds and the guarantee of a fixed market price. He added that in his opinion, the best location would be one and a half miles from here on the river at the village of Monachil, because there is a 20-foot waterfall that would help the water current. Our housing was miserable. I had to lie in the smell of a communal lavatory on a dirty mattress with even filthier pillows. I couldn't sleep because of the bites of the many bedbugs and fleas and when I fell asleep I was periodically woken by the dog lying down on my mattress. There was hardly a glass of water to be had this morning. There was no refreshment because there was absolutely nothing to offer us. I saw the factory, which is water-powered and is so small that it seems impossible that it could process more than $30,000 of paper in a whole year's time, and the profit can't be big, even if you only pay $1500 rent and purchase the rags cheap. We went back hungry and thirsty, I paid 48 for the two horses, we ate together and then separated. The road from here to the factory runs continuously across the slopes of the high hills and through valleys, and below you see the beautiful valley which is watered by a stream that, for lack of something larger, is called a river. I obtained certain information from the Calderon gardener and the bookseller Moreno and learned that a sugar beet factory would be a very bad enterprise as the workers would never agree to grow sugar beets because they suck the fertility out of the soil and they would no longer be able to grow the vegetables that they're used to. Even if they were able to provide the necessary volume of sugar beet, they still wouldn't be able to [70] compete with cane sugar, which is produced on the coast in such large quantities that it's sold for 4rs per arroba, and the lack of firewood would be the final undoing of such a risky endeavor. I like Nottbeck; he is honorable and so honest that he thinks others couldn't possibly be lying and believes everything they say. Of course, he's been hoodwinked many times.

Malaga Tuesday 4 October

I left Granada by carriage on the 1st of October at 6 in the evening. I had a seat which I paid 70 rs for. There were four of us passengers. A very pretty girl of 17 years, born in Valencia, sat next to me,. At first she seemed to have a good background but later I realized from the freedom of her gestures that she had to be a whore. I felt sorry for this poor child who probably ended up in this unfortunate life because of her family or through a vile seducer. She was very beautiful with an aquiline nose, a small mouth and black hair. Her cheeks, which had been rouged, confirmed my suspicion about her social station. When she could not sleep in her corner because of the rocking coach, I took her in my arms and she slept for several hours with her head on my shoulder, which made me so randy that I couldn't sleep. Weakened by my sweet burden I finally arrived on the outskirts of Malaga last Sunday the 2nd, where the little one's new madam was waiting with a coach to take her to S. Antonio No. 13 or 17. I had planned to spend the night with her, but when I heard tell that many were expecting her arrival and that at least a dozen would throw themselves on her that very day, without letting her rest, and when a traveling companion told me that he had already given proof [71] of his love twice, I was scared and restrained myself. Whilst I was gone, the temperature hadn't become more agreeable and it seems to have become even warmer. Unfortunately, there is no steamer until tomorrow evening, when I will leave with the 'Egypcien' bound for Alicante, passage $13. On the promenade of Alameda on Sunday evening, I wanted to sit down on a chair when a gentleman sitting nearby said that the chairs were spoken for, but he invited me to take a seat until his daughters arrived. When he heard that I was Russian, he wanted to get to know me better and he introduced me to his daughters, of whom the one is 13 and the other about 20 years old. I had a friendly talk with them and on their invitation, I promised that I would go visit them yesterday, which I did in good spirit, as I was curious to see how families live here. I found them taking their meal and I suggested that I reserve a box at the theater and come back when they'd finished eating. "Marvelous", replied the eldest daughter, Cristela. They happily agreed. I came back with the ticket and while the girls changed dress, the father and I went to a café. Shortly after we went to the theater together. I gave the oldest daughter my arm. The box was in the second class section, the ticket 30rs and 16 and I wasn't pleased. I was embarrassed that I bought this and sorry I had not paid a duro more and chosen first class. The performance was My *Two* Wives[221] with song and good music and I believe the girls were pleased. The theater is small and hardly fits 400 people. The admittance costs 4 to 5rs; the chairs in the pit about half a duro. After the performance, they escorted me to my dwellingaccommodation. Never had I imagined that Spaniards, [72] who have the pretentions of decent people could live as miserably as this family lives. It seems that they have but two very small rooms, one downstairs and the other upstairs, and in the downstairs room an Italian locksmith also has his work place. The girls care for the household, do all the dirty work of a maid and are therefore not very clean, as their hands and nails show. Their intelligence is barely

221 Olona, D. Luis 1855 Mis dos Mujeres. Madrid.

at the same level as their tasks, and evidenced by the answers they gave, they hadn't the slightest education. But the lack thereof is superseded by their grace and their loving eyes. When I asked if the youngest daughter still went to school, the father answered that there was a teacher who tutored her at home. He couldn't let her go to school because she was getting chased by boys her age eager to have their way with her. He added that the passions here burn hotter and stronger in both sexes. Tonight, there is music on the Alameda. I didn't find my friends there and will go to their house tomorrow.

Thursday 6 October

At Almería on board the French steamer Egypcien en route from Malaga to Alicante.

Yesterday I said goodbye to Mr Matas and his girls who joyously welcomed me, as they thought I'd already departed without bidding farewell. They let me see their tiny rooms and some beautiful perfumed artificial flowers.[222] I bought $5 worth of flowers and gave them to the girls, saying to keep them in memory of me. They offered to send a portrait of the eldest daughter to me in Paris. I hurried to dissuade them of such a silly idea. I told them that their faces would forever remain engraved in my soul and that I didn't want this because of the dangers of the journey. I left them perfectly content in the hope that I would return in the winter. I had to fulfill various formalities [73] linked to my passport, because the French steamer's shipping company wouldn't release my ticket before my passport was countersigned by the health service, who in turn wouldn't sign before it was signed by the police. Since I assumed they wouldn't want to sign until I'd secured the signature of the Russian consul, I went first to the consul, whose name is Rey, and whose agent in Petersburg is Ponfick; then I had it countersigned by the French consul, then the police and finally, by the health service. I took my ticket, for which I paid $13, and after dinner, I boarded. The facilities on board are particularly awful; there are bad rooms and even worse beds for 12 passengers and since there are 15 of us, we had to sleep in threes on benches. Everything here is nasty and you wouldn't think you're on a French steamer.

Alicante Friday 7 October

I'm very pleased with my journey from Malaga to here. I've rarely enjoyed a trip as much as this. I had the very good company of young people from Tenerife who went to school in Madrid. Among them was the Marquis of Florida, a youth of 22 years of age, a man with a profound knowledge of geography and history, whose conversation I relished. He advised and recommended that should I go to the island of Tenerife, to visit his mother in Santa Cruz or at her villa de la Orotaña, which I shall certainly do. The good company I had was compensation for inconveniences on the steamer, which was designed more for a convoy than for passengers. There was no cabin for me to be had and I had to use the cushions of the benches to lie on. We arrived in Alicante this morning at 6, but we had to wait

---

222 Schliemann writes: flores artificiales mezcladas con glándulas – artificial flowers mixed with glands. We suspect that he meant musk or another glandular secretion that is used in perfume.

until 9 to disembark because the government, in its concern for the public welfare of Alicante, had declared us "contaminated". But I couldn't rhyme this with the fact that we had to hoist the quarantine flag and that they made it difficult for us to disembark. They registered the luggage at customs and we departed to the Vapor Inn. I dressed and they gave me an elaborate breakfast, the best in Spain I've had so far. Shortly after I went to the home of Mr Paricio de Satrustague, to inquire of his mother about the good fortune of their youngest son Miguel. Paricio de S. is partner to Mr Lopez of Havana and director of the postal steamers based here in Spain. [74] He is known as a man of great understanding and it's said that he's good at his business. Only the youngest daughter Bernarda was at home, a young maiden around 30 years of age, who welcomed me gracefully. I liked her very much, because she showed modesty and kindness, sincerity and a high character. Contrary to what her brother Joaquin had told me in San Sebastián, she said that Miguel married a poor young American in Havana and that he had established himself in Portland with a Spaniard who had given them money. After his departure from Sonora, she maintained a mysterious silence about the reason and I noticed some embarrassment on her face. He had traveled around Mexico and was appointed English vice consul in Cardenas near Havana; at the age of 22 he had acquired this post because he was born in England and therefore an English national, but he only held the post for two years. Her sister Ventura was with her mother in Madrid, from where they should return tomorrow morning at 10. I also met her sister Inez, who is married to Perfecto Emanuel de Olalde, secretary of the civil government and now residing in Malaga. This lady seemed even simpler than Bernarda. They are beautifully housed, for Alicante in any case.

Barcelona Monday 10 October 1859

I bade farewell to Alicante on Friday at 5.45 pm and went by train to Almansa, from there about 7 miles by carriage, and then again by train. We rode through a beautiful landscape, especially close to Valencia. There is water everywhere and the plain is rich with vegetables, palm groves, orange trees, cherries and other fruit trees.

We arrived in Valencia at 9 am on Saturday. I disembarked at the Inn el Cid. I left to visit the city that is so different from the other cities of Spain, and the Valencian nature is different from that of other peoples' too. The streets are narrow and many are not paved. [75] The houses are generally large and well built. The nature of its peoples is more like the Danish than the Spanish and I found not one beautiful woman. At first sight one notices that the women attend more to housekeeping than to the decoration of their bodies. In Valencia, the neglect and sloppiness you see in Andalusia is nowhere to be seen. It is said that there's a whorehouse here where the girls are examined by doctors twice daily; the government, however, offers no guarantee.

I left at 2 o'clock in the afternoon with the steamboat Mercurio destination Barcelona, where we arrived yesterday afternoon at 1 o'clock. We had to wait two hours before we were allowed to disembark. They checked my luggage again. I took a boy who brought the luggage to the hostel Oriente, where I got a room on the street side, which is the beautiful boulevard called Rambla.

It was Sunday and the streets were full of Sunday walkers. I like this city, which rivals Paris in beauty and solidity of construction. Here, everything revolves around profits and earnings and the houses are built high for more income. From above, the houses are flat to dry the white laundry on the terraces. There is another pedestrian street that is 700 varas long and 60 wide. It starts in the garden of the general, which is very beautiful and full of orange and pepper trees, which blossom and früit simultaneously. The paths are beautiful.

The population seems to me purely gothic; in any case, I do not recognize any trace of the Arab or Romance race here. With rare exceptions, the women are very ugly here. Yesterday I was in the most important theatre, which is much larger than ours, and tonight I was in the Liceo, which they say is even bigger than the Scala in Milan. They performed 'Il Trovatore', unfortunately the play. All the chairs are covered with red velvet, only it has no gloss; the decorations are rude and the actors, actresses and singers are far from great talents. [76] During the break I had a very nice conversation with a lady named Chasq who was sat behind me. She seemed fairly educated for a Spanish woman and prettier than what I've seen here so far. I do not like the aristocracy of the nobility, but I like the aristocracy of the beautiful Spanish women.

This morning I took on a guide who led me everywhere. Don't search for antiquities here. The working and enterprising nature of the Catalans doesn't go well with ancient monuments, and the age-old buildings are demolished to make way for new and practical buildings. The only thing here that might interest an

Figure 22: Barcelona

archaeologist [an antiquarian] is the archive of Aragon, which my curiosity made me enter. There are more than 600 papal decrees, written on rolls of parchment and with large seals. Furthermore there is a huge number of royal letters and other writings, dating back to the year 1000, everything well summarized in books and bound in pigskin. As I was leaving, I encountered the inspector of the archives, Antonio de Bofarull[223], who accompanied me to the museum, where there is nothing but a few damaged statues. Walking along the streets, we met a friend of his named Juan Guitara, who publishes a newspaper here. He brought me two whole large pages from his friendship book so I would immortalize myself in it with a story about a pipe and smoking in 16 languages. Because the request was excessive, I wrote nothing. The streets of Barcelona are very well paved with cobblestones.

On board the steamer Barcelona off San Sebastián Tuesday 11 October

I left yesterday at 1 o'clock in the afternoon with the paddleboat steamer Barcelona. It was very painful to say goodbye to Spain, where I had spent almost two months very pleasantly. There are about 15 passengers, amongst whom a young Spaniard named Emanuel Augustin Calvo, [77] born in Havana, who is traveling to Antwerp to attend school. He'll remain there for a year and then go into business with his uncle in Seville, who is unmarried and has no closer heirs. We have a favorable wind and make good progress, even though we carry a lot of cargo.

Petersburg in the land of Moscow, Sunday 15 November

I returned from my last great voyage and thought of the Tsar and of winter in Paris. At Rothschild's I met Makabit, who had invited me to come to Petersburg. But there was a thief there against whom I had brought a lawsuit. I lost the case and was consequently not in a hurry to return home.[224]

How are your family, your wife and your children? I wrote to you from Rome, after which I visited Alexandria in the land of Egypt. In Cairo I hired a boat with a crew of eight, including a captain, a translator and a labourer. I sailed up the Nile as far as the second cataract and then returned to Egypt. I saw several countries during the trip. I bought a horse and rejoined my fellow-travellers. We travelled cross-country on 13 horses, through the desert to Jerusalem.

223 Antonio de Bofarull (1821-1892).

224 The lost lawsuits doubtlessly refer to the merchant Stepan Solovieff who refused to pay back a loan. Schliemann had hoped to get his money back through summary proceedings in the Commercial Court of St Petersburg, but Solovieff did not accept the ruling of this court and appealed to the Senate. The lawsuit continued and Schliemann did not got his money back until the end of 1863.

Maaike van Asch and Wout Arentzen

The transcription

The text between ↓↓ is written above the line in the handwritten manuscript. Words and passages that Schliemann had deleted are not included in the transcript. Numbers given on the left of the page indicate the page number in the original.

Figure 23: Cover of the diary

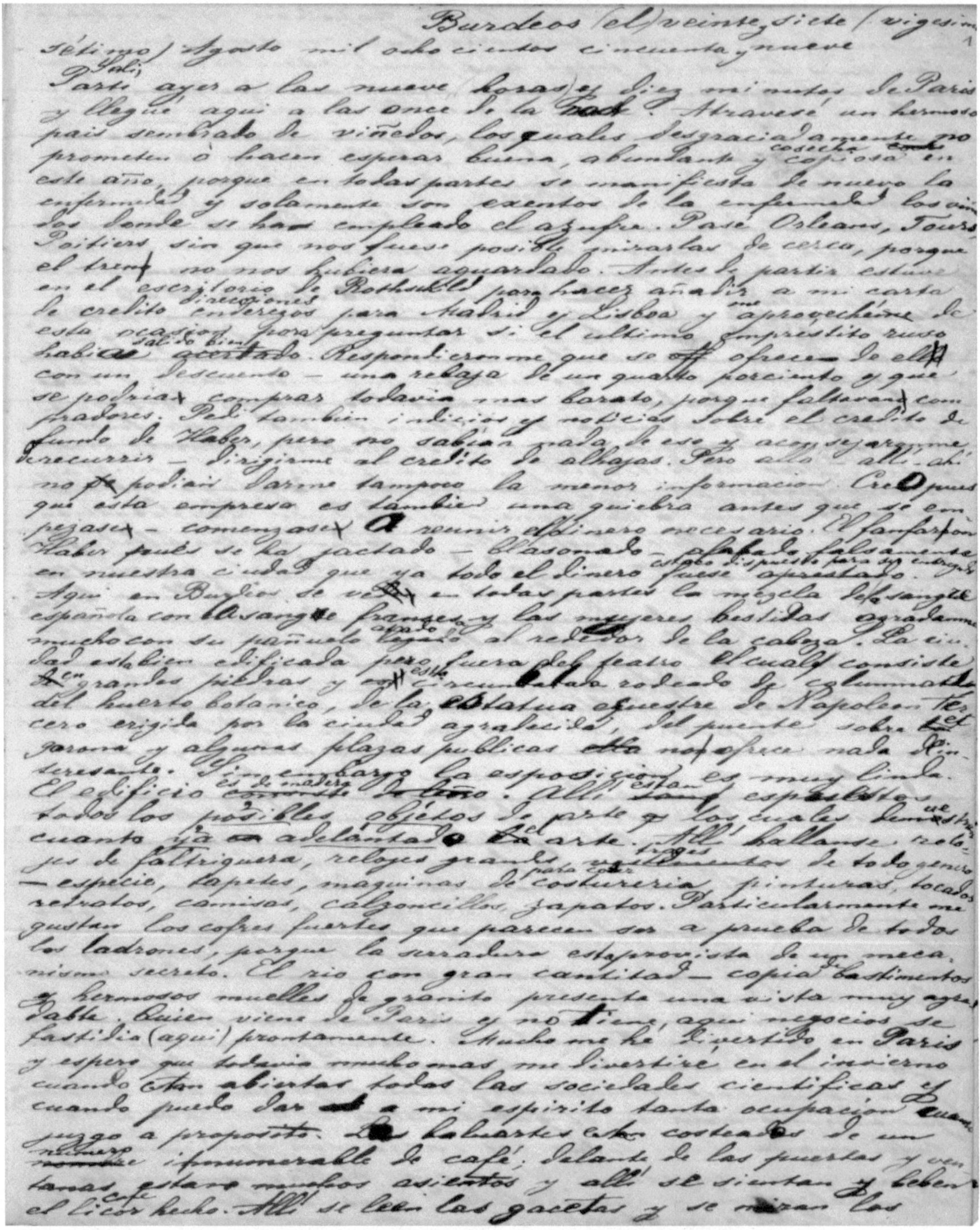

Burdeos (el) veinte y siete (vigesim
sétimo) Agosto mil ocho cientos cincuenta y nueve
Partí (Salí) ayer a las nueve horas y diez minutos de Paris
y llegué aqui a las once de la noche. Atravesé un hermoso
pais sembrado de viñedos, los quales desgraciadamente no
prometen ó hacen esperar buena, abundante y copiosa cosecha en
este año, porque en todas partes se manifiesta de nuevo la
enfermedad y solamente son exentos de la enfermedad los viñe-
dos donde se han empleado el azufre. Pasé Orleans, Tours
Poitiers, sin que nos fuese posible mirarlas de cerca, porque
el tren no nos hubiera aguardado. Antes de partir estuve
en el escritorio de Rothschild para hacer añadir a mi carta
de credito direcciones enderezos para Madrid y Lisboa y me aproveché de
esta ocasion para preguntar si el ultimo emprestito ruso
habia salido bien acertado. Respondieronme que se ofrecen de ello
con un descuento – una rebaja de un quarto porciento y que
se podrian comprar todavia mas barato, porque faltaran com-
pradores. Pedí tambien indicios y noticias sobre el credito de
fundo de Haber, pero no sabian nada de eso y aconsejaronme
de recurrir – dirigirme al credito de alhajas. Pero allo – allí ahí
no podian darme tampoco la menor informacion. Creo pues
que esta empresa es tambien una quiebra antes que se em-
pezase – comenzase a reunir el dinero necesario. El fanfarron
Haber pues se ha jactado – blasonado – alabado falsamente
en nuestra ciudad que ya todo el dinero estaba dispuesto para ser entregado fuese apreslado.
Aqui en Burdeos se ve en todas partes la mezcla de la sangre
española con la sangre francesa y las mujeres bestidas agradan
mucho con su pañuelo atado ligado al rededor de la cabeza. La ciu-
dad esta bien edificada pero fuera del teatro el cual consiste
en grandes piedras y esta circumbalado rodeado de columnata
del huerto botanico, de la estatua equestre de Napoleon ter-
cero erigida por la ciudad agradecida, del puente sobre el
Garona y algunas plazas publicas esta no ofrece nada de in-
teresante. Sin embargo la esposicion es muy linda.
El edificio es de madera consiste de lona. Allí estan espuestos
todos los posibles objetos de arte y los cuales demuestran
cuanto ya adelantado el arte. Allí hallanse relo-
jes de faltriquera, reloges grandes, trajes vestimentos de todo genero
– especie, tapetes, maquinas de para coser costureria, pinturas, tocados
retratos, camisas, calzoncillos, zapatos. Particularmente me
gustan los cofres fuertes que parecen ser a prueba de todos
los ladrones, porque la serradura esta provista de un meca-
nismo secreto. El rio con gran cantidad – copia de bastimentos
y hermosos muelles de granito presenta una vista muy agra-
dable. Quien viene de Paris y no tiene aqui negocios se
fastidia (aqui) prontamente. Mucho me he divertido en Paris
y espero que todavia mucho mas me divertiré en el invierno
cuando estan abiertas todas las sociedades cientificas y
cuando puedo dar a mi espiritu tanta ocupacion cuanto
juzgo a proposito. Los baluartes estan costeados de un
número innumerable de café; delante de las puertas y ven-
tanas estan muchos asientos y allí se sientan y beben
el licor de café hecho. Allí se leen las gacetas y se miran los

Figure 24: Copy of page 1 of the diary

[1]

Burdeos (el) veinte y siete (vigesimo
sétimo) Agosto mil ochocientos cincuenta y nueve

Parti – ↓sali↓ ayer a las nueve horas y diez minutos de Paris
y llegué aqui a las once de la noche. Atravesé un hermoso
pais sembrado de viñedos, los quales desgraciadamente no
prometen o hacen esperar buena, abundante y copiosa ↓cosecha↓ en
este año, porque en todas partes se manifiesta de nuevo la
enfermedad y solamente son exentos de la enfermedad los viños
dos donde se ha empleado el azufre. Pasé Orleans, Tours,
Poitiers sin que nos fuese posible mirarlas de cerca, porque
el tren no nos hubiera aguardado. Antes de partir estuve
en el escritorio de Rothschild para hacer añadir a mi carta
de credito ↓direcciones↓ enderezos para Madrid y Lisboa y ↓me↓ aproveché de
esta ocasion para preguntar si el ultimo empréstito ruso
habia ↓salido bien↓. Respondieronme que se ofrece de el
con un descuento – una rebaja de un quarto porciento y que
se podria comprar todavia mas barato, porque faltavan com-
pradores. Pedi tambien indicios y noticias sobre el credito de
fundo de Haber, pero no sabían nada de eso y aconsejaronme
de recurrir – dirigirme al credito de alhajas. Pero allá – allí – ahí
no podian darme tampoco la menor informacion. Creo pues
que esta empresa es tambien una quiebra antes que se em-
pezase – comenzase a reunir el dinero necesario. El fanfarron
Haber pues se ha jactado – blasonado – alabado falsamente
en nuestra ciudad que ya todo el dinero fuese ↓estado dispuesto para ser entregado↓.
Aqui en Burdeos se ve en todas partes la mezcla de la sangre
española con la sangre francesa y las mujeres bestidas agradanme
mucho con su pañuelo ↓atado↓ legado al rededor de la cabeza. La ciu-
dad esta bien edificada pero, fuera del teatro el cual consiste
en grandes piedras y ↓esta↓ circundado rodeado de columnas
del huerto botanico, de la estatua equestre de Napoleon ter-
cero erigida por la ciudad agradecida, del puente sobre el
garona y algunas plazas publicas, no ofrece nada in-
teresante. Sin embargo la esposicion es muy linda.
El edificio es de madera. Allí estan espuestos
todos los posibles[2] objetos[1] de parte los cuales demuestren
cuanto ya[2] adelantado[1] el arte. Allí hallanse relo-
jes de faltriquera, relojes grandes, trages de todo genero
– especie, tapetes, maquinas de costureria ↓para coser↓, pinturas, tocados,
retratos, camisas, calzoncillos, zapatos. Particularmente me
gustan los cofres fuertes que parecen ser a prueba de todos
los ladrones, porque la serradura esta provista de un meca-
nismo secreto. El rio con gran cantidad – copia de bastimentos
y hermosos muelles de granito presente una vista muy agra-

dable. Quien viene de Paris y no tiene aqui negocios se fastidia (aqui) prontamente. Mucho me he divertido en Paris y espero que todavia mucho mas me divertiré en el invierno cuando estan abiertas todas las sociedades cientificas y cuando puedo dar a mi espirito tanta ocupacion cuando juzgo a proposito. Las baluartes estan costeadas de un numero innumerable de café; delante de las puertas y ventanas estan muchos asientos y allí se sientan y beben el licor café hecho. Allí se leen las gacetas y se miran los
[2]
pasajeros caminantes. La fiesta napoleon y la entrada en Paris de las tropas las cuales han combatido en Italia eran muy lindas. El fuego artificial escedia todo lo que habia visto antes. Se yo tuviese mujer seria muy feliz de vivir siempre en Paris, porque alli se puede vivir como se quiere y enteramente sin fatiga ni pena. Nadie se ocupa de nadie entra en cuidado del otro y si se pone sobre la cabeza un saco ó un sombrero de paja de gran precio todo es indiferente. Yo creia hallar aqui inmensas cuevas ↓bodegas↓ de vino, pero me engañé, por lo menos no he visto nada hasta ahora. Por la vecindad y cercania del mar el clima es aqui mucho menos calido que en Paris y creo que es aqui tambien mas sano.

San Sebastian 28 Agosto 1859

He salido hoy a las seis de la mañana de Burdeos y he llegado a la una a Bayona. El pais que he atravesado es muy esteril y consiste casi exclusivamente en arena. Raramente se veia pequeños campos ↓trabajados↓ laborados y sembrados con un poco de grano. Se dice que toda la tierra arenosa es afluvial del mar hasta Dax donde de repente la tierra se vuelve (se hace) mas fertil. Dax se distingue por una fuente caliente. Bayona es una fortaleza (plaza fuerte) sobre la playa del mar Deseando ver a la pequeña ciudad de San Sebastian tomé mi billete hasta ella y ademas otro billete para el viaje de Bayona a madrid para el cual pagué 140 ½ francos, porque me habian dicho que no halleria billete en S.S.B. No puedo decir cuanto me alegro y cuanto estoy[2] contento[1] y satisfecho de hallarme and España en medio de una gente afable y pronto a servir a los estrangeros ↓a cuanto↓ alcanzan los esfuerzos – ↓sus medios↓. en verdad jamas vi nada de mas agradable y amable que las caras de las criadas y de la Señorita en la fonda donde vivo (me hé ↓hospedado↓detenido). Ellas son muy atentas corteses y estoy enamorado de todos los que tengo vistos aqui hasta ahora. Las calles son estrechas siendo la ciudad una fortaleza rodeada – circundada de fosos – zanjas y bastiones. Fuera de la ciudad hay muchos huertos con higueras y muchos otros arboles frutales. Sobre varios manzanos y peros he visto junto flores y frutos maduros. Cuando pasamos – atravesamos el paseo fuera de la ciudad levantóse – alzóse justamente un aeronauta con su balon ↓globo↓. El hombre se habia ↓suspendido↓ enganchado a las cuerdas bajo el balon y hizo continuamente revueltos – revolteaba. El globo no subió muy alto, porque seria aqui peligroso por motivo de la vecindad (cercania) del mar. Jovenes de ambos sexos bailaron y parecieron alegrarse de la vida. No puedo pasar San Sebastian sin pedir a la Señora Francisca de Satrustegui noticias de sus hijos que hé conocido en tiempo pasado and California y pues yo me hice acompañar de un mozo a casa de la Señora que vive – en el piso alto de una casa and la calle mayor. Casi todas las casas son aqui iguales de altura y construccion. Yo creia hallar aqui mucha desmoralizacion y inmoralidad pero me hé engañado. El magistrado y el ayuntamiento son muy severos y

[3]

hechan fuera de la ciudad qualquier mujer cometa adulterio ↓prostiticion [escandalo↓.]

Por consiguiente no hay aqui putas.

El miercoles 19/31 Agosto en Madrid.

Despues de un viaje muy desagradable de dos noches y 2 ½ días, durante los cuales yo estaba siempre empacado – empaquetado – embanastado como un arenque entre otros muchos pasajeros quienes occupaban los pocos asientos de la diligencia, llegué en fin hoy por la noche al anochecerse a la capital de España.

El primer dia tuve un asiento detras en medio de otros dos y el espacio era tan estrecho que no sabiamos – supimos donde poner los pies – piernas. No puedo sin embargo quejarme de los pasajeros mismos, porque eran muy afables y corteses. Solamente estaba muy asombrado de la conversacion de las Señoras quienes hablaban de las mas indecentes cosas y empleaban tales palabras que yo tuve verguenza de oirlas hablar. La vega que atravesamos es de muy mala mala calidad; en verdad todo es arcilla ligera, arena o peñas – peñascos, los cuales se levantan frecuentemente hasta – a una gran altura. El ferro-carril que ahora se construya desde la frontera a madrid encuentra pues grandes dificultades, por que en todas partes hay peñascos ↓que hay que↓ agujerear, montañas que ahondar – ahuecar y abismos que llenar o juntar por puentes o viaductos. Los pueblos que atravesamos eran todos bien construidos y no se veia huella – vestigio ninguno de la pessima construccion de nuestras aldeas y ciudades de provincia.

El Jueves 20/1 Setiembre. Escribí hoy al Sr. Doré Wunderly 146 calle de Rivoli a Paris: Muy Señor ↓mio↓ y amigo! Vm [= vuesa merced] ha tenido la bondad de prometerme de dirigirme al cuidado de los S[rs] W y Bauer las cartas que pudieran llegar para mi and la fonda del Louvre – Paris. Mucho me obligaria Vm si quisiese preguntar tambien si ya ha llegado mi bagaje o si ya han recibido una advertencia – aviso que el bagaje ha llegado a la aduana. Sirvase Vm rogar que guarden mi bagaje hasta mi buelta. Mucho me gusta la villa de madrid. No del todo no es ciudad de comercio y sera dificil para Vm creer que no hay aqui ningun cambista. Pero tengo aqui carta de credito – estoy aqui accreditado y no carezco de dinero. Menos feliz era en el viaje, porque creyendo que pudiera cambiar oro ruso and San Sebastian y no queriendo perder demasiado sobre el cambio del oro ruso and oro frances y de este en moneda española habia cambiado en Bayona no mas de lo necesario para el asiento en la diligencia. A San Sebastian gaste mis ultimos maravedises y a pesar de los treinta pedazos de oro ruso que tenia en el bolsillo debia ayunar – someterme al ayuno de S. Sebastian hasta Madrid un viaje de dos noches y dos y media dias. Prensado y molestado por la hambre ofrecia frequentemente los imperiales a razon de un duro aunque su precio sea 4¼ $. Pero nadie queria tomarlas y la eterna respuesta era: nunca he visto semejantes monedas, pues deben ser falsas y no daria

[4]

un cuarto por ellas. Muy bonito ha de ser Madrid en el invierno cuando todos vuelven del campo y ↓de los↓ baños de mar, pero ahora ofrece un aspecto desierto despoblado y los calores del otoño cotribuyen a hacer la morada aqui desagradable a aquellos, cuyos medios les permiten de gozar del aire fresco del mar. Recorri hoy varias partes de la villa y gozaba mucho de la vista de los rostros – caras del sexo feminino. en verdad nada de mas gracioso que una española. Ellas estan vestidas por la mayor parte de trages negros y traen – llevan sobre la cabeza ↓velos o mantillas↓ cabestrillos – bandas [negras. El color del rostro indica claramente se son na-] cidas bajo el sol ardiente del Sur. Por la mayor parte las mujeres aqui son mucho mas robustas y provistas de mucho mas grandes senos ↓pechos mas abultados↓ que en nuestro pais. Desearia mucho [hallar aqui una muchacha que quisiera accompañarme en] el viaje del Sur y pagaria bien a quien quisiera recomendarme una. Pero por mas que viaje nunca me atreveré a entrar en una casa publica, tanto mas que del miedo y por el uso de los condones indispensables el carajo no se tendria tieso, mas facil obtendria lo que busco se alquilase cuartos particulares y publicase que tengo nececidad de una criada. Entonces vendrian muchas para ofrecer sus servicios y pudiera escoger quien me guste mas. Pero para eso debiera detenerme mas ↓de lo que↓ deseo y por consiguiente renuncio por esta vez a esta dulcisima empresa. Mañana por la mañana a las once iré a la embajada rusa para obtener de el los permisos para ver gratis – gratuitamente los sitios reales que la capital encierra. Despues solicitaré algún dinero en el banco de Hy P. La construccion de las casas aqui es semejante a la de Paris, solamente es aqui menos solida, lo que facilmente se puede escusar por el grande calor. Las fondas no pueden ser comparadas a las de otras partes de Europa. and verdad les falta mucho para estar al nivel de las peores de Paris. No obstante estoy satisfecho porque el acordarme que en mi vida hé tenido mucho peores me consuela. mucho hé <u>siempre</u>[2] <u>estado</u>[1] equivocado y engañado sobre el caracter de los españoles, a quienes ↓sospechaba demasiado↓ orgullosos para admitir estranjeros and su sociedad o aun para relacionarse con ellos. Al contrario aqui cada una me trata con una afabilidad y cada uno me agasaja – me hace un agasajo que me encanta. Nunca puedo cesar de dar gracias a Dios que me da la posibilidad de poder viajar. Aqui en madrid imitan el ejemplo del emperador de Francia y se hacen en todas partes mejoras. Por ejemplo sobre la plaza a la cual hace frente la fonda en que vivo. Derriban muchas casas y construyen otras para ensanchar la plaza y hacerla mas regular.

[5]

Conté hoy a mi meastro de la lengua española las menudencias – menudos pormenores de mi ascension del volcan Vesuvio. Podia ver al volcan siempre de mis ventanas y magnifico era el espectaculo que se me ofrecia despues del amanecer, porque, aunque ya habian cesado las erupciones del Vesuvio hacia seis meses la lava todavia no se habia enfriado y el monte parecia circundado de rios y arroyos de llamas que se movian lentamente.

El tercero dia despues de mi llegada a napoles (cuando ya habia recorrido la Rusia, Suecia, Dinamarca, Alemania, Suiza, Lombardia, Venecia,

Cerdeña, Romania etc) andúve por el ferro carril
hasta la estacion antes de Pompeo cuyo nombre
no me acuerdo. Allí alquilé – ↓tomé↓ un guia y un
caballo un verdadero Rozinante que hubiera estado
muy agradecido si yo le hubiese llevado – a cuestas.
Ademas me accompañaron dos hombres
vestidos de hombres de bien, que llevaban
corchetes y una litera – silla
de manos. Sospechaba su designio – que
no me accompañaban sino para ayudarme
a subir el volcan y como habia oido que
hasta las mujeres suben sin ↓ayuda↓ – asistencia, yo preferia
subir solo y les esplicaba que no les necesitaba.
El camino nos conducia por de pronto por medio
de viñedos y huertos plantados de higueras y
era tan malo y ↓estaba↓ cubierto de piedras echadas
atropelladamente que con pena podiamos adelantar
– progresar al paso. A mi pergunta porque no
construyen camino empedrado respondio – contesto
el guia que las erupciones del volcan y las
frequentes barrascas – tempestades – tor-
mentas destruyen – arruinan el empedrado. Des-
pues de haber pasado las plantaciones recorrimos
campos inmensos de lava ya fria y despues
algunos arroyos de lava ardiendo, cerca de las cuales
estaban algunos hombres cuya sola ocupacion
era el sacar fuera de los arroyos ardiendos
con largas perebas un poco de lava de ponerla
[6]
en forma de una guirlanda, de introducir en esta
una moneda de cobre que pedian a los pasajeros
estrangeros y de ganar asi una propina. Pasamos
el ermitaño y ya no estaba lejos hasta el pie del
cono, pero el camino es fatigoso y penoso porque
continuamente hay que atravesar las cenizas echadas
en siglos pasados. Dejamos el caballo atado alli y
empezamos la subida – ascension, que era sumamente
dificil, porque el declive es muy escarpado y esta
cubierto a una distancia de mil pies de cenizas
en las cuales se ahonda hasta las rodillas despues
se sube sobre pedazos de lava de los cuales el cono
esta amontonado. La lava resbala bajo los
pies, se arriesga caer atras por un solo falso[2] paso[1].

Por fin llegámos encima sobre el cono y estuvimos
bastante pagados por el trabajo y la fatiga por la in-

mensa y magnifica vista que se nos ofreció en todas
las direcciones. Pompeo pareció estar a nuestros pies.

El humo no nos impedia porque el viento le soplaba
en la direccion de la isla de Capri. La cima del monte
esta cubierta de fuentes o manantiales de agua ardiendo
las cuales estan frequentemente cubiertas de una
corteza delgada de azufre y si no se quiere hacerse
despellejar – desollar se debe estar muy circunspecto.
Acercamonos al crater, del cual salia un humo espeso.
Pero el crater no me pareció bastante grande para
ser digno de un Vesuvio. Pregunté pues al guia
donde estaba el otro, porque ya sospechaba que
hubiese alli dos. Negó, insistí en que me lo mostrase
no quizó, amenazé que no pagaria nada se
no consentia en dar conmigo la vuelta de los cra-
teres y mediante la promesa de añadirle un ↓algo↓ poco
a la suma fijada le empeñé a guiarme. Pero sin
advertirlo nos vimos envueltos por el humo. en vano
nos apresuramos a salir; ya estaba el humo tan
espeso que no podiamos ver la mano delante
de los ojos; perdimos el camino nos Arrastramos por
tierra sobre todos los cuatro miembros. Vagueamos – erramos
dos horas asi sobre las delgadas cortezas ↓capas↓ de azufre
que nos separaban de las fuentes de agua ardiendo
el cual cocia – ardia bajo nosotros y parecia estar
[7]
pronto para engullirnos – tragarnos. El humo
de azufre sufocante hacia dificil la respiracion
y ya creiamos ahogarnos a cada instante. Por fin
volcan tuvo compasion de nosotros y oyó nues-
tras suplicas. Nos teniamos juntos por la voz.
De repente el guia gritó somos salvados y saliamos
del humo. Asi como tres vezes nos acercámos
a los crateres cuyos abismos sentiamos con las
manos. Distinctamente oimos abajo como los
herreros – fabros batian – el hierro
sobre el yunque con sus martillos inmensos
como volcan soplaba el fuego con sus fuelles
Hallámos el lazeroni llorando, porque ya nos
creia perdidos y lloraba que se habia sometido
en vano a la moléstia y perdida de llevar
los comestibles sobre el monte. Saltó pues para
cocer los huevos en una de las fuentes de aqua
ardiendo y estuvieron prontos en un momento.
Regaté – regateaba mucho con el el precio

del almuerzo y lo recibí por fin por cinco carlinos
mientras que pidió al principio ocho vezes mas.
No conocia la fuerza del vino, de que me empeñó
a beber demasiado la alegria de verme salvado.

Descendí pues el declive escarpado, perdi bien
pronto el equilibrio y rodé abajo cabeza arriba
abajo por medio de los pedazos de lava y las cenizas.
mil veces hubiera podido romperme el cuello
pero Dios me protegió y no llevé – me hice ningun mal
ademas de una llaga pequeña al pulgar. Recorri
la Sicilia y me embarqué en Messina con direccion
a Alejandria.

Madrid Viernes 21/2 Setiembre No me gustaba el
quarto and el cual yo estaba colodado and casa de
Viscayna y rogué pues que me diesen otro.
Hicieron lo y mudé de ↓cuarto↓ – aloyamiento. La razon
por la cual queria mudar era que en mi cuarto
anterior era demasiado obscuro. Aunque sienta
frequentemente cosquillas no creo que haya aqui
chinches, pulgas o piojos, al menos hasta ahora
no hé podido coger ninguno de estos parasitos
en frangante delito. Hoy me fui a la embajada
rusa vecina a la puerta and que entramos viniendo
[8]
de Bayona. Pero hacia el camino inutilmente
porque la portera de la casa me dijo que la
chancilleria estaba cerca de la plaza del Oriente
Me fui – me apresuré pues de ir alli. Me dijeron
que no nececitaba hacer poner el visto – ↓visarlo↓ hacer firmar
el pasaporte estando ya visado para España y el
Portugal. Yo pregunté que debia hacer para obtener
el ingreso – el pase en los sitios reales me respondieron
que no necesitaba mas que mostrar mi pasa-
porte. Entonces me fui ↓a la oficina – el despacho↓ en el contador – la escala de
comercio – mostrador de Weissweiler y Bauer y pedí cien
duros a cuenta de (a cargo de mi carta de credito)
me contestaron que el credito era and francos y que no
por consiguiente no podian pagarme sino en esta
moneda. Deducieron la comision y pagaronme
1880r o 94$ and papel moneda. Pedí oro pero no
lo tenian y me dijeron que podia cambiar en
todas partes el papel moneda en oro con un medio
porciento de prima. Les conté que poco faltó para que
muriese en el camino desde Bayona aqui aunque
tuviese en la faltriquera treinta pedazos de oro ruso

que ofrecia a razon de un Solo duro cada uno aunque
valgan 4¼ $. Sonrieronse mucho. Salí y encontre
por casualidad en la calle al abogado mariano
Demetrio de Ortiz en cuya sociedad habia hecho el
pasaje – la travesia de San Sebastian aqui. Andúve
↓j' ai↓ con el a un café y prometióme introducirme maña-
na por la noche a una casa privada de toda segu-
ridad. No dudaba que hallaria aqui la ocasion de
relacionarme con una señorita que consintiera
accompañarme al Sur de España. Hoy hace
mucho calor aun en la sombra y cada uno se ↓todo el que no tiene↓
precision de salir se queda en casa.
queda en casa quien no es forzada de salir.

Apenas hubimos anclado – echado la ancla – ancora
en el puerto de Alejandria cuando acercóse a nues-
tro vapor una barquilla – lanchilla con muchos
arabes que gritaron Srs una propina. Esta fué la
primera y la ultima palabra que oí en el
oriente. Muchas veces cuando subimos el
Nilo con viento favorable con gran velocidad
nos gritaban den Vms una propina como se
debesemos de hacer escala – hubiesemos de dete-
[9]
nernos para darles lo que pidiesen. Muchas veces
se echaban – arrojaban al agua y nadaban
hacia nuestro buque al cual se enganchaban y no
dejaban hasta que no recibian el presente deseado.

El viento Shirocco, que es humedo en Italia y
destruye los monumentos de la antiguedad con
rapidez es muy seco en Africa, porque toma so-
lamente sobre el mar la humedad. en Egipto no
llueve nunca y la sequedad y la aridez del clima
preserva los monumentos de modo que por ejemplo
las pinturas y los dibujos hechos hace 4000 años
parecen tan frescos como si fuesen hechos ayer.
La pesadumbre que debia someterme voluntariamente
al engaño de un solo judio mozo de cordel me empeñó
a estudiar la lengua arabe con tanto esmero que
acerté a hablarla en algunas semanas. Las casas
arabes son mal construidas y se componen de cuatro
troncos o cepos de palmeras hundidas en el suelo y
llenos de lodo del nilo. El techo consiste en ramas
y hojas de palmeras puestos encima de los muros
improvisados. Las puertas son tan pequeñas que hay que
arrastrarse por tierra para entrar. en el

interior todo el ajuar de casa – amue-
blaje consiste en una estera de junco, sobre
la cual duerme toda la familia. Los trages consisten
en algunos harapos – andrajos sucios. Hasta la
edad de seis años las mujeres van enteramente
desnudas y llevan solamente una angosta cintura
de cuero sobre el coño – las portas. Una vez oí gritos formida-
bles saliendo de una casa en Corusco. Entré and
ella para ver la causa de tanto ruido. Vi dentro
dos hombres que tenian sobre la mesa un niño
recien nacido de sexo feminino. La madre armada
de una aguja con hilo estaba occupada en coser
el coño de la niña. Pregunté la causa de tan dolo-
rosa operacion. Me contestaron que era solamente
para impedir la prostitucion y para preservar la
virginidad hasta el matrimonio. me explicaron
que despues de la operacion acuestan la niña
sobre el estomago hasta que cure – sane la parte
agujereada – horadada.

Este noche dí una vuelta sobre el
magnifico paseo llamado el Prado
[10]
el cual estaba estorbado – lleno – embarazado de gente
de todas clases y Estaba lleno de gozo porque
nada de mas agradable puedo imaginarme que
de hallarme y pasearme en medio de las her-
mosas españolas. Cuando nos sentimos felices
entonces el tiempo pasa con velocidad y sen-
tia que la tarde fuese corta. De repente
mi alegria se mudo – se torno en pesadumbre, porque
me vino a la memoria la familia F. con la
cual me hé relacionado en Oriente y a la
cual he prometido venir a verla a Canada
sino me impiden circunstancias que estan encima
sobre los calculos y esfuerzos humanos. Resolvi
a escribir al padre la carta siguiente: Cuando
me despedí de Vm en Jerusalem y Vm espresó
su pena que yo, impedido por otro empeño,
– que yo, siendo ya ↓comprometido mi palabra↓ con otros – no
pudiese – pueda hacer con Vm el viaje por la
Siria, respondí: La divina Providencia se ha
servido arreglarlo ↓disponerlo↓ asi y hallaremos despues que
era asi mejor. en verdad apenas hube llegado
a Betlehem cuando recorriendo en mi memoria
mi vida pasada hallé en ella razones que

me parecieron demasiados podersas para ↓establecerme↓ colocarme
en Canada. La naturaleza nos ha asi constitui-
do que hallamos alivio y consuelo de nuestras
penas en el comunicar a otros las cosas que
oprimen nuestros corazones. Me apresuré pues
a confesar al Sr Smittan las razones que
segun mi parecer me han de impedir de
colocarme en Canada y la razon que me atrae
allí y que me hace esperar que seria allí el
mas feliz de los mortales. en medio
de los peñascos del desierto y en medio del inmenso
calor sobre la orilla del mar muerto en to-
das partes discutia – deliberaba yo con S. sobre
dicho punto que formaba el unico objeto de
mis pensamientos cuando velaba y
de mis sueños cuando dormia. Por fin con-
venimos ambos que el hombre que
[11]
quiere ser digno de la carrera – que Dios
le ha indicado por este mundo debe antes sacri-
ficar la felicidad de su vida de manchar –
ensuciar su conciencia o desviarse de sus prici-
pios y maximas. Si Vm no hubiese rehusado
mi compañia cuando yo la ofreci a Vm en
Cairo para el viaje a Jerusalem; o si despues
de haber puesto mi orgullo a tan dura prueba
a Cairo Vm me hubiera dicho una palabra
a gaza, donde le repetí a Vm tres veces en
purisima lengua inglesa que habia hecho mi con-
trata solamente hasta Jerusalem y que seria allí
libre de asociarme a cualquiera otra compañia
entonces con grandisima alegria hubiera
acompañado Vm no solamente por toda
la Siria sino tambien sin duda ninguna hasta
Canada, porque sincera amistad para Vm
y su familia y la alegria que hallaba en su
sociedad me hubiera hecho imposible separarme
de Vms. De vuelta a Jerusalem yo estaba fir-
memente resuelto a perder
antes mi vida que a emigrar a Canada. Una
profunda tristeza y una pesadumbre como no hé
sentido nunca se apoderaron de mi. ↓Con semejante disposicion↓ de espiritu. Hubiera
sido un compañero fastidiosisimo para mis
socios. me separé pues de ellos y recorri
sola toda la Siria con un dragoman y dos criados

y mi grande alegria era de plantar mi tienda
en el mismo lugar donde, segun las indicaciones
de los indigenos Vms habian tenido las suyas
pocos dias antes. en nablous dormi and casa del
mismo judio, en la cual Vms habian
tomado su refugio cuando Vms llegaron alla
mojados por la lluvia. en Tiberias hallé los
ultimos vestigios de Vms. Estando acostum-
brado a vivir siempre and el gran mundo
y siempre en compañia la soledad me fastidiaba
mucho y no pudiendo ya sostenerla – padecerla
mas me resolvi a ir directamente
a grecia sin haber visitado el Libanon
y Damasco. Con esta determinacion llegué
[12]
a Sidon y fui al consulado frances para
informarme acerca de la salida – partida
de los vapores franceses. La casualidad quízo *F* que
encontrase allí a mis buenos amigos los dos condes
italianos en cuya compañia habia hecho el viaje des-
de Cairo hasta Jerusalem. De ellos aprendí que
Vms habian partido ↓marchado↓ la vispera para Constantinopla.
A Beirut encontré al Sr Sacerdote Taylor el compañe-
ro de Vm, que me díjo *F* que Vms irian de Constan-
tinopla inmediatamente a grecia deseando la Seño-
rita Mary mucho – teniendo la Señorita M. muchos
↓deseos↓ de ver a aquel país. Si ↓yo↓ pues ↓hubiera marchado↓
para Grecia hubiera ↓debido↓ necesariamente
encontrar allí a Vms, que no debo mas ver
porque me hubiera costado la vida si hubiese
hecho – emprendido con Vms el viaje por la grecia.
Me resolvi pues a tomar la direccion opuesta y me
fui a los montes del Libano, recorrí las nieves eter-
nas para ver a los cedros contemporaneos de Salomon,
fui a visitar Balbeck y Damasco y me embarqué
en Beirut dos semanas despues de su ↓marcha↓ – partida
de Vms me detuve de intento algunos dias en las
islas griegas y and Smirna y temblando de todo el
cuerpo de miedo de encontrarles en Atenas lle-
gué por fin allí. Seria trabajo inutil probar
a dar a Vm una debil idea de lo que padecí – sufrí
debiendo a toda fuerza separarme de Vms que quiero
y estimo ↓por↓ encima de – sobre todo and este mundo –
en verdad ↓mis↓ sufrimientos – padecimientos eran
mayores que cada descripcion ↓ponderacion↓. La afliccion ↓continua↓ me enervó

hasta tal ↓punto↓ grado que caí múy enfermo en Atenas, una
fiebre nerviosa me arrostró a la – la cama
– poco a poco perdí toda mi fuerza y me hice
tan debil que no podia mover ningun
dedo; horas enteras me hallaba desmayado
y a cada instante aguardaba la muerte.
Cuando estaba en tal desesperado estado me
trajeron cartas de mi banquero en San Peters-
burgo. Estas cartas escritas en lengua
sueca ninguno en la fonda donde yo estaba
alojado podia leermelas y yo mismo
[13]
estaba demasiado debil para poder leer. El medico
se fue pues a la embajada sueca y suplicó al
secretario de la embajada que me les leyese.
Este vino a mi cama y me las leyó. El tenor
de dichas cartas producío sobre mi el efecto de la
almizcle, porque mi banquero me dio parte que cierto
individuo de nombre Solovieff, que me debia 18 meses
ha 17000£ y que, no hallandose en estado de pagar
habia obtenido de mi la gracia de pagar solamente la
1/5 parte and dinero y de darme
por el resto letras de cambio a un año, 2, 3 y 4
años no solamente no habia pagado las leteras
de cambio del priméro de los quatro ↓plazos↓ terminas,
sino que habia presentado
contra mi una queja en la cual habia atacado mi
honra. Lleno de furia yo mandé que me trajesen
inmediatamente en un coche y que me llevasen –
conduciesen al vapor frances de las mensajerias
imperiales que debia salir aquel dia en direccion
a Constantinopla. El medico se opúso luego ↓d' abord↓ pero
dejandome ↓el casi↓ consintió. Dejé pues
la gratificazion para sus servicios y me hice
llevar al Piraeo, el puerto de Atenas, donde me
hice llevar sobre el vapor y poner sobre el puente
El aire fresco del mar me curó – sanó pronto
y despues de haber estado algunos dias en Constan-
tinopla ya podia marchar. Me fui al embocadero
del Danubio, subi este rio hasta Pest, recorri la
Alemania hasta Stettin, de donde el vapor ruso me
me conducío en tres dias a San Petersburgo.
No tenia ninguna necesidad de volver alli para
obtener el pago ↓cobro↓ de las leteras de
cambio, porque and nuestro pais como

en todas partes hay una ley que prescribe que
toda letera de cambio deba ser pagada cuyo firmante
no declara and el espacio de tres dias despues
de haberla firmado que no ha recibido la valor
de ella. Volví pues solamente para salvar mi honor
y llegado a S.P. deposité todas las leteras
de cambio de S. and el tribunal de comercio
confié la tan facil defensa al mas habil abo-
gado y partí por Paris, porque tuve miedo
que mi salud quebrantada por la fiebre no pudiera
ostener ↓la rabia – desazon↓ que me deba naturalmente
[14]
causar el pleito. and Paris me quedé 12 dias y asisté
a la fiestas de napoleon. Depues me fui por
Burdeos y los pirineos a España, and cuya capital
me hallo desde algunos dias. me quedaré todavia
algun tiempo aqui para visitar despues granada,
malaga, Sevilla y Cadiz. Despues visitaré el
Portugal y pienso pasar el invierno alternativamente
en Paris, Roma y Napoles. La vispera de mi salida
de Jerusalem tuve el gozo de visitar con Vms los se-
pulcros de los reyes, la plaza donde creció el arbol
cuyo tronco dió – la santa cruz y por fin
las canteras de que sacaron las piedras para la
construccion del templo de Salomon. Cité entonces a
Vms el celebre discurso que el infeliz Kosuth ha
pronunciado en el capitolio de Washington en Julio
1851 y and el cual entre otras ↓cosas↓ dice: "la palmera crece
mejor bajo un estupendo peso". No pude entonces ↓contener↓
me reir a carcajada tendida de estas palabras tan
estrañas. Pero la dura y amarga esperiencia me ha
enseñado despues cuan verdadero es el sentido de
ellas. en la alegria y el gozo del corazon frequentemente
↓nos↓ olvidamos de ↓dar gracias a↓ Diós por sus favores; nunca
lo olvidamos and la pesadumbre y la afliccion. Al con-
trario la afliccion del corazon empeña al hombre
a cogerse con ardiente celo a los principios y
maximas de la honradez y de la virtud, que
dan realce al – que realzen el alma y la enoblecen.
En la alegria nos cuesta poco para suprimir los remor-
dimientos de la conciencia, en la afliccion antes quer-
iamos mil veces morir que desviarnos del camino
de la virtud. El retrato de Vm y de su familia
siempre presente en mi alma sera mi orgullo
durante esta vida y mi caro y fiel ta-

lisman and las horas de la tentacion. Por todas
las riquezas del mundo no cambiaria la ↓memoria↓
de Vms. Cuan ↓consoladora↓ y sublime
es en la afliccion y la pesadumbre del corazon
la creencia en un otro ↓mundo↓ mas feliz, cuan
celeste ↓es↓ para mi la esperanza ↓de↓ que volveré a
ver a Vms en aquel mundo y viviré quizás
↓entonces↓ junto con Vms and los vinculos de estrecha
amistad durante la eternidad en
comparacion con la cual el resto de nuestra
[15]
vida no es sino un momento. Animo pues
corazon afligado! animo and la creencia!
Rogando a Vm de presentar mis salutaciones
cordiales a su cara esposa, su hija y su hijo
y repetiendo a Vm que nunca podria olvi
dar de encerrar a Vm y su familia en mis súplicas
a Diós quedome con sentimientos de reverencia
y amor amigo suo
Q.R.S.M.
Hy S

En tiempos pasados
El Dinero era el unico objeto de mis pen-
samientos y Mammon no tenia nunca un
adorador mas ferviente que yo. Ahora me
he hecho enteramente indiferente al dinero,
porque no estando acostumbrado a los excesos
no pienso que pueda nunca gastar la mi-
tad de mis rentas. Vm me dijó en gaza que
en su ciudad podria facilmente tener 10% com-
prando propiedades. Con gozo le man-
daria a Vm mi dinero para colocarlo alli
pero de esta modo entablaria una corresponden-
cia con Vm, lo que no me siento la fuerza de
hacer, porque cada carta suya me abri-
ria de nuevo las llagas sangrientes del cora-
zon. Espero nunca volverme a Petersburgo
porque hace allí demasiado frio y las 3 ½ sema-
nas que pasé allí en Julio yo me senté en
mi casa de campo envuelto en ↓mi↓ capa de
peles. Pero mi nombre esta allí bien conocido
y si Vm quisiese escribirme sencillamente
H.S. en S.Pbg su carta me seria mandada
inmediatamente y ella llegaria siempre pos-
ticalmente a mí, porque mis agentes saben

siempre donde mis cartas me alcanzarian.

Despues de la partida de Vms fue hallado en su cuarto un pedazo cortado de un periodico conteniendo lo siguiente: el reverendo T. nos favoreció con su visita la semana pasada pero parecido – semejante a un cometa se marchó dejando detras una inmensa cola de luz.

[16]

Madrid Sabado el 22/3 Setiembre. Tengo la costumbre de lavarme por la mañana todo el cuerpo con aqua frio. Apenas hube hecho esto hoy por la mañana cuando ya mi maestro topó – chocó – empujó – llamó a la puerta. Dije: entre Vm y el gran hombre apareció. Fuimos a dar una vuelta al gran huerto ↓al buen retiro↓ al cual llegamos despues de haver atravesado el Prado. Creo que cuesta mucho de mantener aquel huerto en buen estado y regar regularmente todas las plantas, porque el suelo es arenoso y el sol ardiente abrasa – quema todo. El huerto se distingue por la inmensa cantidad – copia – abundancia de hermosas flores y por algunos estanques bastante grandes, profundos y hermosos, sobre los cuales nadan patos o anades que, acostumbrados a ser nutridos por los que vienen, se apresuraron acercarse a nosotros en la esperanza probablemente de que echaramos a ellos ↓migas de pan↓ pedazos de ↓cortadas de↓ pan. Como la isla de Cuba and gran parte contribuye a la prosperidad de España, me parece que esta ultima debria hacer todo ↓lo posible↓ para conservarlas Todos saben que el presidente actual de los Estados Unidos fúe elegido por no otra razon que por la de que esta en favor del anexion de Cuba, y sin duda alguna se Buchanan hubiese sido presidente durante la guerra de Crimea, cuando la Inglaterra y la Francia estaban demasiado occupados para pensar en intervenir – en una guerra en las antillas, entonces sin la menor duda el hubiera hecho una tentativa de tomar a Cuba ↓para apoderarse de C.↓ Pues, como yo quiero mucho a España los valientes habitantes de ella no me pueden tomar de mala parte, si, preocupado de la futura suerte de Cuba que tanto les importa yo les aconsejo de mandar allí un tan

gran ejercito como solamente puedan jun-
tar – porque entonces los americanos
se aprovecharan de la oportunidad de las
[17]
circunstancias. Tomando en mano el mapa
geografico ↓y mirando por el hieroscopo de los pueblos↓
profetizo que grandes mutaciones se
haran de aqui en cinco años. El poeta griego
Homero, que vivio 1000 años antes de la era
cristiana dice: ἔσσεται ἦμαρ ὅτε ἄν ποτ᾽ wλώλῃ Ἴλιος
ἱρὴ ναὶ λαὸς ἐϋμελοίο Πριάμοιο lo cual sig-
nifica en buen castellano: El dia vendrá cuan-
do perecerá el Santo Ilio y el pueblo del bien
vestido Priamo. De la misma manera digo: vendrá
el dia y aquel dia no estalejos cuando la flota
rusa y la flota francesa juntas destruiran y
haran estallar ↓blow to atoms↓ la flota inglesa hasta aqui invencible
y cuando la orgullosa Ingleterra sera reducida
a un estado mas debil que el chiquito reino de
la Holanda. Canada entonces[2] tendra[1] verguenza de
pertenecer a la gran Bretaña
se anegará a los
soberanos estados de america septentrional y
brillara sobre la bandera de las estrellas.
Filibusterismo
americano conquistara ↓entonces↓ Jamaica casi sin
resistencia de la parte de los perezosos negros.
Cuba la perla de las Antillas sera hecha território
americano luego que estallara
la guerra entre Rusia, Francia y Inglaterra. la
efusion de sangre será inmensa porque los españo-
les les defenderan su isla heroicamente. Cuando
estalle la guerra contra Inglaterra los ingleses
suplicaran en vano a las otras naciones que
les ayuden, nadie vendra a su secorro, porque
cada uno odia Inglaterra y verá con gozo
su caida. Gibraltar sera entonces vuelto a
España a la cual pertenece segun las
leyes de la naturaleza; la Rusia ↓moldavia y valachia↓ tomara
la Turquia européa con exception de la
Albania, macedonia i Tesalonia que seran
vueltas a la grecia. La Francia
se apropiará las islas ionicas y se divi-
dira con Rusia Asia minor, la Siria
y Palestina para estender allí la civilisacion

[18]
reprimida por el barbarismo ingles.
El Egipto se hara un estado neutral y independiente. Las Indias orientales seran fraternalmente divididas entre la Francia y la Rusia; – la Australia se proclamará republica. Victor Emanuel rey de Cerdeña y la Lombardia quedara el vasallo de napoleon.
Para no hacerlo demasiado poderoso los ducados de Toscana, modena, Parma y Romaña (parte del Estado Pontificio) seran unidos en un ducado separado. El dominio ↓clerical↓ de los sacerdotes continuará todavia a Roma y el destierro del sumo pontifico a Jerusalem no sera hecho mas ↓que en↓ los proximos ↓siquientes↓ 5 años. Poniendo el hieroscopo un poco mas lejos y mirando los acontecimientos de 1865 hasta 1870 veo que la Francia toma toda la Italia con escepcion de la Cerdeña que deja a su vasallo. Vm no debe tomar ↓en mal sentido↓ – no ha de tomar de mala parte lo que el odio me inspira a decir de los ingleses. Mi odio es limitado a los ingleses y no se estiende a los sencillos escoseses cuya franqueza, lealtad afabilidad y ↓esprit comun↓ inteligencia natural siempre me ha encantado. Asi como tenemos la costumbre de pintarnos – dibujarnos en la imaginacion un dibujo de los objetos desconocidos que nos interesan asi me hé pintado tambien en el espiritu montreal.
Debe ser una hermosa y limpia ciudad muy semejante a Filadelfia (aunque mucho menos grande) con un escelente puerto y un pueblo de mucho espiritu comun que junta al espiritu emprendador americano, la prudencia y solidez escocesa, de que conté a Vms en mesa redonda en Jerusalem un glorioso ejemplo en los aldeanos de Kirkaldy, de quienes el uno se hizo ↓en 1739↓ Visir turco y el otro marescal ruso.
Una breve descripcion de nuestro viaje de Jerusalem a Petra Vm hallará and el periodico
[19]
Times del 27 de Mayo. Por desgracia la España ha siempre sido visitada por viajeros quienes no conociendo ninguna palabra de la lengua

no podian ser recibidos en conversacion y quienes
pues se fastidiaban. Ellos naturalmente no
podian proporcionarse los informes necesarios
y salian de España mas estupidos que habian
venido. No obstante se pusieron a escribir libros
sobre la España, libros llenos de las mas abomi-
nables mentiras. Como cosa natural contaban que
↓habian sido↓ atacados y robados and las mismas calles de
madrid. Por eso tienen en todas partes muy falsas
↓ideas↓ de España. Le aseguro a Vm
que nunca en mi vida he sido recibido mejor que
aqui y por eso no tengo nada mas que hacer que
de tratar a la gente con un poco de orgullo pero al
mismo tiempo con una franqueza leal. Un conocido
mio que ha vivido aqui algun tiempo me asegura
que estaba aqui estorbado – embarazado de
invitaciones y que hallaba aqui una hospidalidad
como no habia hallado aun en Rusia. El caracter
español es un poco orgulloso, pero este orgullo es
laudable, porque no emana – procede de vanidad
sino del conocimiento de dignidad propia como hombre
↓stoop to indignitino↓ y del menosprecio de aquellos que
se bajan y se humillan para obtener ventajas tem-
porarias. Como natural consecuencia
del clima ardiendo el español es mas vivo que
el frances pero su vivacidad cadenciado por el
orgullo da – presta a todos los movimientos de su
cuerpo una gracia imposible de describir. Esa gracia
que parece natural es realzada por la naturaleza, que
aficiona los contrastes y esta lujosa aqui and el
producir (por la mezcla de la sangre Saracena,
romana, cartaginesa y de otras naciones que
han sucesivamente habitado esta tierra) mara-
villas de belleza de ambos sexos. La belleza fe-
minina aqui esceda todo lo que hé visto hasta ahora
Como consecuencia natural del clima ardiente
las mujeres maduran y marchitan aqui mas
pronto que bajo las zonas mas setentrionales,
pero les queda siempre la indes-
cribible gracia de los movimientos del cuerpo.
[20]
la grande vivacidad junta con
tanta dignidad, ellas hablan con tan leal franqueza
y juntan ↓a todo eso↓ en edad adelantada tanta esperiencia
porque el orgullo nacional ↓y el patriotismo↓ obliga aqui a cada uno

a ocuparse del bienestar del pais y no hay criada
por ↓mal↓ educada que ↓sea que↓ no sepa todos los pormenores
de las discusiones en las camaras de los cortes, no
hay cochero que sentado en el pescante de su coche
↓los faroles del coche farolas↓
↓no↓ trague ↓devore↓ por la noche a la luz de las linternas hacia
las cuales se agobia ↓inclina↓ – se cuerva los discursos de las
Cortes y todas las demas novedades que los periodicos
ofrecen) que uno pudiera enamorarse de una españo-
la de sesenta cinco años; al menos no puede
↓uno menos de↓ mirarla sino con profundo interes y respeto.
La lengua castellana es tan sonora ↓y ademas tan rica↓ y armónica
que en comparacion con ella todas las demas quince
idiomas que hablo me parecen feas. Convencidos
de la belleza de su lengua maternal y demasiado ocu-
pados and otras ciencias que puedan
contribuir al mejoramiento de la condicion de la patria
los españoles no estudian otros idiomas y por con-
siguiente es necesario que cada estranjero que quiere
gozar de su viaje and España y enriquecerse
aqui de informes utiles debe, ante toda otra cosa, tener
un conocimiento perfecto de la lengua. El cielo
ardiendo de España es causa que las pasiones estan
aqui mas ardiendo y fervientes que and paises mas
boreales y las leyes de la naturaleza no parecen
ser en armonia con las leyes de los hombres que
han instituido aqui la monogamia. Tambien
no se puede negar que esta ultima ley,
aunque mantenida ↓nominal↓ y en la forma, no lo es en realidad
y que la sociedad española se considera tener
de la naturaleza el derecho de considerarse ↓mas o menos↓ eman-
cipada de dicha ley. Por eso se halla aqui mas
libertad entre ambos sexos que en cualquier otro
pais y frecuentemente aun la conversacion
de las Señoritas y Señoras de la alta
clase rueda sobre objetos indecentes ↓obscenas↓ y lleva
tales palabras que mis criadas brutas tendrian
verguenza de cuchichear la una a la otra al
oido. Una semejante[2] emancipacion[1] de las
costumbres reinaba en Roma antigua y en
[21]
grecia antigua como distintamente lo demuestran
los antiguos autores y las pinturas y emblemas
que Vm habra visto and ciertas casas
en Pompeo cerca a Napoles. en

los paises donde el mohametanismo permite la Poliga-
mia, ↓y satisface asi↓ las pasiones de los hombres
las leyes de la naturaleza estan
de acuerdo con las de los hombres y
por consiguiente no se halla allí casi ninguna
prostitucion.
Soné – toqué la campanilla y cuando el
criado víno le dije: traiga – me Vm un cafe
Pero repito a Vm que no quiero café frio como
recibi el otro dia. Si el cafe no es muy caliente
y ardiendo no lo quiero y le reenvio ↓vuelvo a mandar↓. Antes
Vm debe echar – verter un poco de agua ardiendo
en la jicara – taza para calentarla. Tambien
traiga Vm pan y manteca de vaca y sal, mos-
taza, porque no quiero almorzar. Pero, como pienso
irme hoy a la corrida de toros la cual dura-
ra hasta las 6 de la tarde haria ciertamente
mejor de almorzar.
Madrid Lunes 24/5 Setiembre. Ayer habia convidado
un conocido mio a comer conmigo – admitir mi
comida. Es un hombre muy flaco y no puedo ima-
ginarme, que, segun dice, tenga tres mujeres
de las cuales dos fijas y una por caprichio. Des-
pues de haber comido una sopa o menestra
de yerbas y pan, vaca bien grasa – gorda
con legumbres, tocino con habas, salchichas con
guisantes, tomates, judias, carnero asado,
ternera asada, puding, helados- sorbetes con
tortas hojaldres y bizcochos y para los postres
ciruelas, peras, albaricoques, higos y guindas ↓cérise↓
salimos y fuimos a un cafe. Aqui sirven (yosim)
el cafe de una manera grosera, llenan la
jicara de modo que ensucian la mesa y ↓el traje↓ los ves-
tidos del parroquiano y del todo no sirven
la copa ↓le petit verre↓ de coñac. Ademas el cafe es amargo y no me
gusta. Despues dimos una vuelta larga sobre
el paseo de la castellana y entramos en un
baile al aire libre ↓en plein aire↓. Parecio come se todas
[22]
las modistillas de medio pelo de madrid estu-
viesen allí juntas – reunidos. Los bailes me
parecieron diferentes de los nuestros, porque
aqui la mujer pone la cabeza sobre el pecho
del hombre y en verdad parece que la jode bai-
lando. en verdad no se pudiera desear una me-

jor cita con una mujer que en estos bai-
les. Allí habia muchas mas mujeres que caballe-
ros y como cada uno eligió – escogió sola-
mente las bonitas todas las feas quedaron
descuidadas. Dos bandas de música tocaban bastante
bien. El huerto estaba bien iluminado con an-
torchas de todos colores._Despues de haber gozado
de ↓la vista de↓ este baile durante media hora fuimos
al prado y apenas salimos cuando ya vimos
la diferencia entre la clase de gente del baile
y la ↓de la↓ calle. El Prado estaba estorbado – em-
barazado de paseantes. Al fin del paseo un
charlatan ha fundado una tienda ↓casilla↓ de madera
y bajo ↓los redobles↓ el batir del tambor un hombre colorado
y un muchacho muestran al publico las ma-
ravillas que el arte del charlatan artista produce.
Les faltan a ambos los delanteros[2] dientes[1]
toman una dentadura de dientes artificiales ya
preparada para ellos y la ponen and la boca
muestran cuan firme se ↓los↓ tiene y principian
– empezan a mascar pan que ponen en la boca en
largos pedazos. Al mismo tiempo distribuyen
escritos o papeles impresos en los cuales son
descritos las pormenores.

No conseguí hoy a ver el palacio real, porque
lo estan gobernando ya todo el verano. Pero
me entregaron papeletas para visitar la
armeria real donde se guardan sus armaduras
antiguas. Aunque estaba cerrada hoy obtuve
↓el paso↓ mostrando mi pasaporte. Al mirar
las armaduras pesadas no pudo sino lastimarme
de la degeneracion del genero humano, porque
ahora no se acertaria sino con gran
trabajo levantar uno de dichas armaduras
que los antiguos llevaron dias enteros sobre
[23]
el cuerpo._Sin embargo mucho depende del ejercicio
y si and vez de cultivar el espiritu sometieramos
el cuerpo desde la niñez a los egercicios gimnasti-
cos nos hariamos quizas tan fuertes como nues-
tros antecesores – antepasados. La armeria con-
tiene gran numero de sables, espadas, escopetas,
espuelas y otras cosas. and todas partes se advierte
a los visitadores que no deben tocar los objetos.
Fui despues a las Cabellerizas reales, en las cuales

hay un inmenso surtido de hermosos caballos de
todas razas. Allí hay enfermeria, taller y.o.
Tengo miedo de sobrecargar mi estomago, porque creo
que mi tristeza y melancolia proviene – procede
dimana solamente de mi abundancia de sangre como
natural consecuencia del poco movimiento que tengo
y del mucho sustento – alimento que estoy aco-
stumbrado a tomar. Con mucho gusto me accordaré
siempre de mi viaje a Jerusalem aunque probablemente
en esta vida no hare mas aquel viaje. La vida
en las tiendas en el desierto era ciertamente suma-
mente uniforme, a menudo haria que someterme a
grandes privaciones de todo genero. No obstante,
cuando ahora miro atras todo parece como si
lo viese ↓atraves de↓ un velo de color de rosa, porque aquella
deja muchos puntes lucientes y lustrosos al nebuloso
horizonte de la memoria. El camino de Jafa a Je-
rusalem es uno de los peores que hay, aunque sea
frecuentado anualmente por millares de pe-
legrinos. Se dice que el camino era antes mejor y
que ha sido arruinado por los arabes para detener
la marcha del ejercito egipcio. Todo esta sembrado
de piedras de todos calibres y grandaries, de modo
que yendo al paso uno arriesga romperse la
cabeza. Jerusalem no se ve ↓mas↓ que pocos minutos
ante de llegar allí. grande fue mi alegria cuando
por fin vi delante de mi el objeto –
de mi viaje; grande fue mi engaño cuando
entré y recorri las estrechas – angostas tortuosas
sinuosas calles resbaladizas and tal grado que
no se podia ↓montar↓ a caballo sino con gran peligro
de resbalar. El templo o la iglesia del santo
sepulcro se halla and mal estado – mala condicion
[24]
porque poco a poco se ha destruido el techo de la cu-
pola de manera que la lluvia cae ahora sin
piedad y lastima sobre las cabezas de los pios.
La competencia y el mutuo odio irreconciliable
entre las tres religiones dominantes es la razon
que la santa iglesia amenaza ya a caer
en ruinas. Si las reparaciones costasen milliones
cada una de las sectas se apresuraria a hacer
lo necesario, pero la una no permite a la otra
la honra de hacer las reparaciones y como no
reina entre ellas ni concordia ni union ni acuerdo

la iglesia no sera en pocos años mas que un monton
– una pila de ruinas. La reverencia con que el
viajero entra and la santa ciudad se torna and indig-
nacion cuando atentamente observe el lugar
en el cual la santa iglesia que
contiene a la vez el santo sepulcro y el monte
Calvario donde nuestro Señor fue crucificado,
porque es imposible que el uno y el otro – lo uno
y lo otro hubiese estado en medio de la ciudad,
la cual no puede haber sido mas grande que
ahora por estar circondada de altos y escarpados
peñascos. Tampoco es posible que el sepulcro
hubiese estado tan vecino al lugar de la cruci-
ficacion. La ley de los judios les prohibia
enterrar los muertos a la distancia de cien pasos
de la ciudad. and medio de la iglesia se halla
una capilla que encierra el sepulcro. La capilla
consiste en dos piezas. en la primera esta
el tronco de la columna, a la cual Jesus fue atado
cuando le fue puesto sobre la cabeza la corona
de espinas y cuando el fue azotado. en la se-
gunda esta el sepulcro, pero el marmol de ella
me parece demasiado blanco para poder creer
que este ataud pudiese ser el verdadero. en
la segunda pieza se halla siempre un fraile que
da a los piadosos que entran para hacer sus
súplicas a Dios una de las flores que estan
puestos alli en vasos de vidrio y les riega con
agua de Coloña estendiendo al mismo tiempo
la mano para recibir la propina. Este comercio
[25]
en lugar tan santo me disgusta tambien.
El Sr Oriz ↓vino a buscarme↓ fue por mi hoy a las cuatro y
fuimos juntos a la Corrida de toros; el edificio
en que se hacen las corridas
consiste en un gran anfiteatro ↓en el↓ que caben 17/m
hombres. Era un espectaculo curioso de ver en
las calles con que ansia, aprieto y premura
los hombres se apresuraban a ir a la corrida de
toros. Sobre la cara de cada uno estaba pin-
tado el miedo de llegar demasiado tarde. en
todas partes desde la puerta del Sol hasta las
corridas estaban plantados estacionados municipales a
caballo para mantener orden entre los carua-
jes; los que iban habian de eseguir en una fila – linea↓ tenerse en una

serie y adelantar a proporcion –
a medida ↓que↓ los primeros llegaban y depositaban
– apeaban sus pasajeros. Los caruajes vacios
se apresuraban a volver con toda priesa para
tomar otros pasajeros. Las diligencias estaban
sitiadas y tomadas por asalto. Teniamos buenos
sitios llamados tabloncillos de grada 1[era]. Es muy buena
institucion que el dinero que se cobra de las
corridas esta ↓dedicado↓ al beneficio de los pobres y de las
enfermerias. Eso aumenta todavia el gusto que
tienen los españoles de hallarse en las corridas.
El pobre prefiere padecer sufrir la hambre o
empeñar su ultimo caldero al monte de piedad (caja
publica al socorro sobre prendas para amontonar
acumular el dinero necesario para ver las corridas
de toros. La prensa era inmenso. Era un espec-
taculo curioso de ver tanta gente junta. Bajo el
tocar ↓los ecos – sonidos↓ de la musica abrieronse la puertes de la
arena y los alguaciles ↓a caballo↓ salieron seguidos de
los toreadores y dos yuntas de mulas, que fueron
empleados despues para arrastrar los caballos
y toros muertos. Los hombres inclinaronse de-
lante del presidente o director de las corridas
los alguaciles salieron, los toreadores tomaron
sus capas en mano el juez dió la llave de
los establos, abrieron la puerta y un toro
furioso salió y echóse sobre los toreadores
[26]
los cuales estendian siempre sus capas delante
de los ojos de los toros y esquivaban – se escapa-
ban despues con mucho destreza saltando sobre
el parapeto. Tres toreadores se tenian siempre
a caballo con largas ↓lanzas↓ provistas con
pequeñas picas. Cuando los toros se arrojaban
sobre ellos y hundian sus inmensos cuernos
en el cuerpo de los caballos entonces ellos/los
toreadores, los picaban con sus picas pero no
les podian hacer sino heridas pequeñas. Cási
a cada ataque de los toros a los caballos
estos caian con sus ginetes, cuyas piernas
estaban envueltas en hierro de modo que no
eran heridos por la caida. Cuando estaban
↓por el suelo↓ (arrastrados por tierra) su posicion era muy
peligrosa, pero entonces los toreadores a pie
venian pronto y los socorrian teniendo los ca-

pas delante de los ojos de los toros. Apenas
hubo un toro hundido sus cuernos en el vientre
de un caballo que ya salieron las entrañas
a menudo salio solamente[2] una al principio[1]
veniendo ella bajo los delanteros[2] pies[1] del
caballo este mismo pisandose las
sacó fuera todos los demas
porque los toreadores no cuidando de los pobres
animales los hacian correr hasta que caian
muertos por la tierra. A menudo los toros
saltaban la vallia – el olivo pero no hacian
daño a nadie. Despues venian los toreadores
con lancetas envueltas en papel and forma
de una pluma y las ponian con mucha
destreza en la ńuca de los toros. Los torea-
dores eran aplaudidos cuando plan-
taban las lancetas ↓banderillas↓ bien y silbados cuando
no acertaban o si las lancetas caian
despues de ser puestas. Varias veces
las envueltas de las lancetas contenian
baquetas que se encendian y circondaban
el atonito toro de chispas. Despues el uno
o el otro de los toreadores armado de su
[27]
espada en una mano y de su capa en la
otra teniendo siempre esta ultima delan-
te de los ojos del toro procuraba de aprove-
charse de un momento favorable para hun-
dir la espada entre los dos huesos en la nuca
del toro y matarle asi. Mucha lastima me
dieron los caballos que debian correr con
la mitad de las entrañas pendientes del
↓vientre↓ estomago. Apenas fue muerto un toro que
ya las yuntas de mulas fueron enganchadas
a los cuellos de las bestias y trotando
bajo los golpes del azóte – látigo
sacaban los cadaveres fuera de la escena.
Esos espectaculos me parecen demasiado cru-
eles y feroces y creo que debo acostumbrarme
a ellos para poder hallarlos hermosos.

Pienso de escribir lo siguiente al Sr D^n Const: Fehlei-
son de Petersburgo, actualmente en el Havre

Muy S^r mio!

Apenas me hube puesto en la diligencia con direccion
a Madrid cuando ya reconocí que los españoles estan

todavia muy atrasados en comparacion con los
demas pueblos porque estabamos seis personas en un
espacio que apenas podia contener tres. en verdad
hubiera deseado que mis piernas fuesen de guta percha.
Hace ya ocho dias que estoy and la capital y confieso que
aun el viajero que viene de Paris puede complacerse
sumamente aqui algun tiempo por la gran dife-
rencia que hay entre las costumbres, el modo de
vivir, las recreaciones publicas por fin bajo todos
respectos. El sol ardiente de España hace que la
sangre arda aqui mas que en otras partes, que cada
uno se siente tiranizado por la indisoluble ley
de los hombres que permite el matrimonio solamente
con una mujer. Tambien segun lo que veo y
oigo no creo que haya en Madrid un solo marido que
sea fiel a su esposa (y pocas esposas fieles a sus
maridos). La prostitucion es aqui mucho mas
general que aun en Paris, pero aqui el amor se
da solamente por capricho y no es venal
como allí. en verdad la castellana es demasiado
orgullosa para vender su amor y ciertamente ella
[28]
no se humilia a eso si no esta forzada por
la mas grande necesidad. Seria superfluo asegurar
a Vm que no hay constancia en los caprichos de
las castellanas y que su amor se rueda
como una veleta. Si digo a Vm que las
mujeres de San Sebastian eran hermosas le puedo
asegurar ahora que la naturaleza esta aqui toda-
via mucho mas lujosa en el producir maravillas
de belleza. La augusta reina de España hace una
rara excepcion de la regla general, porque ella
vive con su augusto marido and los vinculos del
mas ardiente amor cuyas pruebas ya le ha dado
en 8 niños hermosos como botones de rosa.
La aficion a la cosas materiales – al materialismo
transluce mas en las recreaciones ↓diversiones↓ publicas.
Si el gobierno publicase hoy que mas no hubiera
mas corridas de toros, una revolucion terrible estallaria
al instante en toda España; si los bailes publicos
al aire libre estuviesen cerrados solamente una
semana el pueblo los estableceria en las calles pu-
blicas. Como tengo dicho la inclinacion natural del pueblo
esta al amor material, a las recreaciones ruidosas
y estrepitosas. Cuando ayer estuve sentado en un

asiento de la plaza de toros y los picadores
sin piedad ni lastima dieron las
espuelas y hicieron galopar a sus caballos
cuyos vientres ya varias veces estaban
hundidas por los cuernos agudos del toro*F*, al
cual los toreadores a pie habian arrojado en
la nuca gran numero de banderillas o lancetas
llenos de baquetas que ↓si encendieron estallaron y↓ circondaron
el toro furioso de un mar de llamas de modo
que bramaba de dolor y de desesperacion*FF* entonces
no pudo contenerme de esclamar: que atroz es-
pectaculo. Una Señora que estaba sentado cerca
de mi oyendo esas palabras se sonrojó -
de colera y dijo: que mal gusto tiene Vm
de no sentirse trasportado y conmovido viendo
ese sublime espectaculo. en verdad añadió
aqui entre 17/m espectadores no hay ninguno
[29]
tan bruto que no este electrizado de tan
celesta vista. Y en verdad no habia fin
al gozo y a la alegria que se manifestaba
por continuaos aplausos y gritos de bravo
F de manera que las entrañas o ya pendian en grandes
masas fuera del vientre o que eran
sacadas fuera poco a poco por los mismos caballos
que al correr atropellaban con sus delanteros
pies sobre una entraña que pendia.
Las riñas de los gallos no es menos terrible
de ver. Me parece que la Divina providencia
no a concedido al hombre el derecho de atormentar
de semejante manera las pobres bestias que sienten
el dolor tambien como nosotros. Cuando los es-
pañoles se burlan de nosotros que tenemos
en Rusia todavia la esclavitud entonces respondo
En nuestro pais solamente una parte
de los aldeanos son esclavos sin embargo su es-
clavitud cesará ↓dentro↓ de pocos meses mientras vuestra
esclavitud no ↓concluira – acabara↓ nunca, porque vosotros
sois esclavos de vuestras pasiones y esta esclavitud
oprime vosotros mucho mas que el azote del
dueño ha nunca oprimido los al-
deanos.
El segundo dia de la fiesta de pascua
habia siempre en Jerusalem la costumbre que el
Santo fuego era distribuido entre los fieles reunidos

en la iglesia. Se decia que el fuego salia repentinamente del Santo sepulcro. El patriarca griego creyendo que esta costumbre de la ↓edad↓ media no era conveniente para nuestro tiempo de progreso habia abolido aquel engaño. Pero el pueblo insistia con grandes voces a que dicha costumbre fuese reinstituida diciendo que sin ella la fiesta de Pascua no era nada. Asistí a aqel engaño que ordinariamente cuesta la vida a muchas personas porque la ↓muchadumbre↓ es tan grande que muchos son sufocados o atropellodos. Gracias al consul ingles obtuvimos ingreso en una de las galerias de donde pudimos ver el espectaculo. Muchos habian pasado la noche en la iglesia para obtener buenos sitios o al menos uno cual-
[30]
quiera. Me apoyé sobre la balustrada pero bajo el peso de dos ↓gruesos↓ sacerdotes que se apoyaron sobre mi. Otros me prensaron y me apretaron de ambos lados de modo que sudaba como si estuviese en un baño ruso. Despues de haber aguardado asi 3 horas fue representado como si el fuego hubiese salido del sepulcro porque uno de los individuos que se habian acercado a la abertura que hay en la capilla y tenia and ella sus cirios de cera sacó repentinamente la mano fuera con el cirio encendido y encendió algunos cirios de otras personas que a su vez dieron la luz a otras de manera que en pocos instantes toda la iglesia no era mas que un mar de luces mientras que no ↓callaban cesaba↓ la griteria y de alegria. Despues los soldados turcos estacionados por la iglesia para mantener el orden hicieron ↓sitio – lugar↓ para las procesiones de los sacerdotes de las tres religiones dominantes que dieron tres veces la vuelta de la capilla con santas imagines y cruces and las manos. No pude contenerme de pensar lo que Christo diria si repentinamente se alzase – se recuscitase en medio de tales hechos. El corral del templo esta siempre lleno de vendedores ↓mercaderes↓ de cruces, anillos y de todo genero de santas alhajas, que se fabrican en Betlem de nácar. Dos dias antes de mi marcha vi que un judio con alhajas de nácar entró en el aposento de un viajero americano alojado en la misma fonda y que poco despues entró en

el tambien la mujer de mi corazon con su
padre. Entré tambien y habiendo dado a la Señorita varias imagenes la rogué que escogiese
para mi las mejores. Escogió tres de las cuales
la una era una caja cuya tapadera representaba la entrada solemne de nuestro S[r] en Jerusalem montado sobre un asno. La otra eshibia
la Santa cena muy bien hecha y la tercera
representaba otro acontecimiento de la Santa
Escritura pero no me acuerdo el objeto.
Sobre el reverso de cada una de esas alhajas
escribí: a la S[ra] del honorable Sr J.F. con
[31]
los cumplimientos ↓de H.S. y sus↓ mejores deseos para constante
bien estar. Despues empaqueté cada uno separadamente y despues todas juntas en periodicos griegos.
En la pequeña caja↓ cajita↓ habia metido – puesto una carta
↓con muchas dobleses para hacerla caber↓
(que debia plegar ↓ ↓ muchas veces para hacerla entrar)
Del contenido siguinte. Muy S[ra] mia! Ya a bordo
del vapor que nos conducia de Mesina a Alejandria
observé que Vm y su familia
me manifestaron sentimientos de amistad
los cuales saludé con entusiasmo tanto mas cuanto
ví que no estaban apoyados sobre motivos interesados, porque del todo no tenia amigos cuando era
pobre, y estoy convencido que todos los amigos, que
logré cuando la guerra de la Crimea tan llena de
acontecimientos, aumentó seis véces la fortuna
con que volvi de la California en 1852
me dejarian se cesase de tratarlas con comidas
de vino de champaña o de darles negocios en que
↓pudieron sacar producto de mi↓ pueden ganar dinero por mi. No es posible para
mi de espresar a Vm cuan estoy obligado a Vm
por su amistad desinteresada. en verdad desde
que conozco a Vm y a su familia el Canada me parece
un paraiso y me estimaria el mas dichoso de
los hombres si pudiera vivir allí por el resto
de mi vida en su vecinidad de Vm ↓cerca de Vms↓. He prometido
a Vm de venir al Canada en este año; pero mil
veces hé sido engañado en mis mas seguras esperanzas y no hubiera debido prometerle cosas futuras, que dependen de la Providencia divina. Los
fragiles dedos del hombre no pueden ni parar, ni
aflojar la rotacion de la rueda de los destinos

humanos! Si pues↓ no me es licita↓ la felicidad de volver
a ver a Vm y a su cara familia entonces me
despido por eso para siempre de Vms asegurandoles
que nunca puedo olvidar a encerrar en mis
súplicas a Dios el bien estar de Vm y de su
familia y que en medio de los huracanes sobre
los oceanos bramantes y en la monotonia
encantadora del valle del nilo, a las orillas
heladas del golfo de Finlandia y sobre las ondas
sonrientes del mediterraneo, en medio de las
diversiones de las sociedades en Paris, Londres,
Petersburgo y Mosca y en las ocupaciones urgentes
[32]
del comercio, en peligros y fatigas, en
trabajos y dificultades – nunca puede pasar
y nunca pasará un solo instante en el que
mi espiritu no sea con Vms cuchicheando
– dando sus bendiciones a Vms
alegrandose de su felicidad
y llorando sobre sus desgracias. Con senti-
mientos de reverencia y estima quedome amigo de Vms.

Mi orgullo estaba muy humillado en el Cairo
por la denegacion de Ud de permitirme a jun-
tarme a su compañia, pero mucho mas todavia
en Gaza cuando vm me dejó sin convidarme
a compañarle desde Jerusalem a Beirout
aunque habia dicho a su esposa de Vm en el Cairo
que habia hecho mi contrata con los dos con-
des italianos solamente hasta Jerusalem por la
razon que esperaba juntarme allí a Vms y
aunque habia dicho a Vm en Gaza dos veces
que en Jerusalem seria libre de juntarme a
cualquier otra compañia y la tercéra vez que
el Sr Smyttan me habia convidado a acom-
pañarle a Petra y que le habia contestado que
reflexionaria sobre ese punto.
Esperaba al menos hallar a mi llegada a Je-
rusalem en la fonda de Hauser un billete
en que Vm me convidase a acompañarle
por la Siria y no hallando ninguno
principié a pensar que Dios queria
darme un aviso de no ir con Vms.

El gran cometa observado en Con

Ya hace mucho tiempo habrá pasado el
estrecho de gibraltar el gran Cometa, que

fue observado en Constantinopla hacia el fin
de Diciembre 1858. Su paso a las
primeras cataractas del Nilo, a Jerusalem,
al Libano y al monte Sinai habra
aumentado todavia su lustro y los astro-
nomos del Canada veran ahora aquel
cometa por muchos años en toda su gloria.
[33]
En frente de mis ventanas en la fonda de la
Vizcaina vive un Señor, cuya criada creatura
amorosa me da a entender que quiere ser
jodida. No sabiendo como hacerle llegar
una papeleta ↓cartita↓ le gritó cuando cerró las cortinas
que viniese a las diez a mi cuarto y no
sé porque no ha venido.
Madrid 26/7 Setiembre 1859. Ayer fui a ver
el museo de pinturas. El surtido es quizas
el mas grande del mundo y allí hay muchas
obras exquisitas – ↓maestras↓ tanto de la escuela española
cuanto de la escuela italiana. Hay tambien
algunas salas con esculpturas que ofrecen
muchas buenas cosas del arte antiguo y moderno.
Alli estan suspendidas varias copias
del genero de adornos de las paredes de la
Alhambra. Los adornos con sus colores vi-
visimos presenetan una vista magnifica.
Nos pusimos en camino y despues de 1 ½
hora llegamos a Betlem, donde hay gran
numero de conventos y casas hermosas de obispos.
Aun las casas de los aldeanos presentan una
vista muy limpia y pueden caber allí unos
4000 habitantes. Las mujeres se distinguen
por su belleza. Entramos en la iglesia
y nos acercamos con reverencia al pesebre
donde nació el hijo de Dios. El pesebre
esta tallado en un peñasco. Continuamos nuestro
camino y llegamos por la tarde a Hebron,
donde a mi aturdimiento y asombro los judios
me hablaron en ruso. Establecimos nuestras tiendas
y fuimos al arbol – una encina – que se dice
encina de Abrahan, pero no me parece tan
vieja. Naturalmente la corteza esta llena de
nombres de viajeros que han querido eternizar
su memoria. en Hebron esta la mezquita and
la que estan enterrados Abrahan, Isaac y Jacob.

Pero siendo los arabes muy fanaticos no me
atrevia entrar. El dia siquiente fuimos al campo
del jefe de los beduinos. El campo consistia en
muchas tiendas miserables. Nuestra inesperada
[34]
llegada esparcia la alegria sobre
todos los rostros. El viejo escogió los hombres que
habian de acompañarnos, me dió un buen
↓caballo padre↓ blanco y principiamos el largo y pe-
ligroso viaje. Pasamos muchos destrozos y ruinas
de antiguas ciudades y entre ellas algunas
bastante grandes. Almorzamos por la mañana
a las cinco o ↓las↓ seis. teniamos tres
tiendas; el almuerzo ya estaba serbido en la tercera.
La mesa estaba cubierta de manteles limpias
huevos (fritos), tortillas y carnero
con sal, pimienta y mostaza formaban nuestro
frugal almuerzo. Pimienta comimos mucho. A las
doce tomamos el segundo almuerzo, que consistia
en tortillas envueltas en papel, carnero asado,
naranjas y confituras. Teniamos con nosotros
cerveza inglesa. La cucina todos los utensilios de
ella, las tiendas, las camas de hierro, los tapetes
colchones – mantelas almohadillas, conssint – almohadas,
colchas – cubrecamas etc debiamos llevar con nosotros
porque ademas de ↓la↓ suciedad no se halla nada en
las tiendas de los beduinos. Estos van siempre con
pies desnudos y no tienen trages; algunos andrajos
– harapos sucios cubren se desnudez en parte. Sobre
la cabeza llevan una especie de gorra para protegerse
de los rayos ardientes del Sol. La primera
noche cuando apenas hubimos establecido nuestras tien-
das y cuando los beduinos apenas se hubieron
sentado al rededor del fuego que acabaron de en-
cender de arbustos, saltó repentinamente el uno de
ellas en el aire con grandes gritos y clamores
por haber sido mordido de serpientes. Pero felizmente
no era asi. Las serpientes habian quierido acercarse
al fuego y se habian introducido bajo el trasero
del hombre. Las mataron y desde entonces
eramos a menudo ↓durante las noches↓
– nos despertabamos a menudo sobresaltados
por los gritos de los beduinos. Es en
verdad una maravilla que no hayamos sido
mordidos durante todo el viaje, porque

[35]
venenosos serpientes y scorpiones se introducían
a menudo en nuestras tiendas. Era tanto mas
peligroso cuando nos debiamos siempre levantar
la noche para aguas fuera de la tienda
porque no teniamos ↓basinillas vasos urinales↓ dė noche. Llegamos
por fin a los peñascos de Petra y empleamos
todo un dia para subirlos, porque era un camino
terrible a menudo ancho apenas de
un pie, resbaladizo y escarpado; era nada
mas que una vereda estrecha que nos conduci
al declive de las peños y nos mostraba con-
tinuamente la muerte porque a cada paso veia-
mos los profundos abismos bajo nosotros y
debiamos cerrar los ojos para no ser victimas
de los vertigos.
Madrid miercoles el 7 Setiembre. Hoy me
fui a la Alameda propiedad del duca[226]
de Ozuna. Obtuve el billete en la contadoria.
Tomé un carruaje porque la Alameda dista
una legua y media. El cochero se paró a la estre-
midad del parque y debé ir – quedia de ir
un quarto de hora a pié. El portero me mos-
tró el interior del palacio. Los quartos son
pequeños pero bien amueblados y todas las pare-
des cubiertas de hermosas frescos o con
tapices de seda y estan llenos de quadros y litografias.
del comedor[2] sobre el parquete[1] del empedrado
con ladrillos o atabones esta imitado con buen
suceso el celebre musaico del museo de napoles
en el cual es representada la batallia
entre Dario y Alejandro el grande. Hay otro
aposento en el cual se halla un billar, otro
con una representacion en madera de una
corrida de toros; un torrero tiene delante del
torro furioso su capa y observa – aguarda con atencion
el momento favorable para clavar –
encajar su espada en su nuca; otro torrero
↓ya ha↓ clavado encajado hundido su espada en el pescuezo
del toro; otro hombre llega con su yunta de
mulas enlazar un nudo al rededor del
pescuezo de un torro matado. Hay una
[36]

226 Italian? Spanish duque.

magnifica alcoba del duca con colchones
ropa blanca – lienzo, colchas – sobre camas
y almohadas, con lavabo de porcelana, ce-
pillos de dientes etc en todos los quartos hay
chimeneas, tenazas. Sin embargo todo me pare-
ció un poco descuidado y en decadencia.
Los mas quartos estan cubiertas de tapetes.
El jardinero me acompañó despues en el jar-
din, que es en verdad una maravilla aqui
sobre este suelo arido. Sin embargo con todo el
cuidado los céspedes son quemados por el sol
ardiente. Las hojas de los arboles se marchi-
can bajo la influencia del sol ardiente. Hay
aqui una magnifica casa de abietes [abejas] cyas
cajas son asi hechas que abriendo una puer-
tilla se ve por el vidrio como trabajan las abi-
etes, como componen la cera y ponen el miel
en las celdas. Se hubiese estado encinto hu-
biera en verdad abortado, porque el
jardinero me hizo entrar and una casita en
que estaba sentado un ermitaño ↓de madera↓ leyendo
su biblia; en una cabaña vecina estaba su
cocina con fogon de madera. Me introdujo
en otra casa donde estaban sentados juntos
en la mesa redonda un viejo con su mujer y su
nipote [nieto]; delante de ellos habia pan, pollos
asados y dos vasos que parecian a medio
llenos de vino pero ellos eran dobles y la ma-
teria roja que en ella habia los hacian parecer
como si estuviese vino. Arriba en la misma
casa estaban quartos. Habia una otra casita
con un pequeño surtido de cañones de cobre
y vecino estaba una pequeña fortaleza rode-
ada de una zanja. Habia un himbernaculo
en el cual se ponen las plantas esoticas o
tropicas durante el invierno. Las naranjas
por ejemplo mimadas por el gran calor no
pueden soportar – sostener el frio. Habia
un pabellon para bailes, algunos estanques
sobre los cuales nadaban cisnes y anádes.
Delante el gran estanque habia un pabellon
[37]
en que los Señores hacen sus preparativos para
la pesca a la caña o con redes – hilitos.
Habia una fuente asi arreglada con pequeñas

peñas y guijarros que no moja el que se sienta
sobre un banco bajo la fuente mientras que
moja a todas que se hallan a alguna distancia.
En un muro estaba colocado una pequeña escultura rota por las manos de algunos visitadores barbaros. El unico hereditario del duca
que no esta casado es su primo el principe
angrona marques de Javal quinto.
El duca se llama Piedro Alcantara
Tellez giron. Ha erigido en el jardin un
monumento a su abuela Maria Josepha
Pimentini duquesa de Orsuna. El administrador de sus haciendas es el marques Alcanizes calla Alcala. Al duca pertenecen
1422 pueblos y 300 ciudades.
Madrid 28/9 Setiembre 1859. Ayer por la mañana
empaqueté mi ropa, pagué mi cuenta y entregando
mi bagaje a los criados para que lo guardasen –
conservasen hasta mi vuelta, me fui con mi maestro
al ferro-carril. Era la fiesta de la Santa Virgen
y habia mucha gente, que queria aprovecharse de este
dia de descanso y de reposo y divertirse and la campaña. Teniamos en nuestro departamento del coche
buena compañia y entre los otros la Señorita
Maldonado hermana de Rafael G. Maldonado
que vive en Madrid Valverde 46 – 3° . mientras
que su dicha hermana es alumna en el Colegio
de Señoritas nobles en Toledo. Mi entregó su tarjeta el hermano y díjo que presentandola en el colegio y diciendo que venia de la parte de esta Señorita
me mostrarian el colegio. en la ciudad de Toledo
de cerca 15/m habitantes la alineacion de las calles
y la construccion de estas respira la media edad,
porque las calles son sinuosas, pequeñas y angostas
mientras que la fachada de las casas y el
arreglo – disposicion interior es grosera. La
catedral que se cree en España ser una de las
mas celebres del mundo por el merito de sus quadros, estatuas y adornos debe ser visitada antes
que se visita Italia, porque todas las principales
[38]
catedrales de esa son mucho superiores en todas
respectos. El altar y el coro estan circondados de
esculturas magnificas de piedra y encierran labores en madera de sumo merito. Las balustradas

son de ferro martillado; en una capilla hay sobre el altar cubierto de manteles de plata una estatua de la Santa virgen con el hijo diós en su brazo de plata pura; las ventanas estan adornadas de magnificas pinturas; las bóvedas de la iglesia son superbias y suntuosas pero toda la iglesia podria ser puesta sobre el piso de la Catedral de San Pedro en Roma y su grandarie desapareciendo en la inmensidad del espacio pareceria muy pequeña, porque 4 o 5 veces ella pudiera caber en la catedral de S. Pedro. Almorzamos carne asado, anguila bullida y fruto y fuimos despues al نقرق o palacio construido por Carlos quinto. Hallandose sobre una colina este palacio presenta una aparencia grandiosa, pero el interior es cási enteramente destruido y los cielos rasos y bóvedas han caido en todas partes. Hay alli una escalera de caracol de gran merito porque bajo de la una escalera de caracol hay una otra que sigue las mismas curvas. El guardia un soldado nos condujo a los cuartos bajos que consisten en caballerizas y caben 25/m caballos; alli hay tambien fogones y cocinas._Fuimos despues a la casa de los locos; hay cerca 40 hembras y otros tanto masclos. Entre las hembras hay una marquesa cuya razon se ha vuelta por el mucho leer novelas. Ella parece tener poco mas o menos de 35 años y respondio a mis preguntas de una manera confusa y poco inteligible; tiene allí su criada. Las mujeres estan separadas de los masclos. Duermen sobre colchones por tierra, tienen lienzo blanco almohadas y colchas. Los cuarto son muy grandes y no hay muebles en ellas. en cada cuarto hay una chimenea guardado por balustradas curvadas de hierro para que los locos no se quemen al fuego._Fuimos despues al colegio de señoritas, donde fui recibido con

[39]

agasajo, porque ya habia hablado de nosotros la señorita Maldonado. La directora o vicaria me condujo and todas partes del vasto edificio fondado por un Cardinal bajo el reino de Felipe II. Las cenizas del Cardinal reposan en la iglesia.
El edificio asemeja mucho a una lonja, porque el medio esta abierto y solamente protegido por una

tela – cubierta contra el sol y la lluvia. Abajo y arriba hay grandes corredores para dar vueltas. Volvimos a Madrid por el tren de 6 ¼. Habia en el coche la S[ra] D[ña] Alfilar Alfaro calle de carretas 15 cuarto segundo con su hija joven bastante viva pero menos guapa de 16 años. Dije a ellas tantas galanterias y lisonjas que ambos jouliano los labios de gozo. Ese genero de doncellas no pregunte si el hombre tiene gibas de detras y de delante – es giboso y corvado, si es viejo o joven, hermoso o feo; todo lo que desean es de estar casadas cuanto antes, porque por desgracia hay aqui mucho mas mujeres que hombres y muchisimo mas mujeres que hombres que puedan alimentar – nutrir una mujer. La madre me dió la direccion de su casa y debia prometerle de ir a verla a mi vuelta de la granja.

Madrid 9 Septiembre 1859

G. Sterky y hijo, St Petersburgo

Muy S[res] mios!

Confirmando a Vms mi ultima carta de Paris con fecha del 12/24 pasado tengo el gusto de acusarles el recibo de su muy agradecida del 14/26 pasado del que anoto que Vms han recibido de vuelta su carta de credito de Rpl 160/m. Anoto tambien la venta de inscripcion de R 500 a 111.
Cuando subira el precio de las acciones del ferro carril Dunaburgo Riga les dare a Vms mis instrucciones acerca de la venta. Me cuestan a mi 200 acciones 3 ¼ premio y mil acciones 4 ¼ premio. Cuando sera anunciado que se puede entregar de nuevo pagamentos sobre las acciones de dicho ferro carril entonces sirvanse Vms entregar prontamente los pagamentos sobre mis acciones, y si Vms no tienen entonces
[40]
dinero de mi sirvanse Vm discontar una parte de las cartas de cambio que Vms habrán tomado para mi en discuento para proporcionarse los fondos necesarios._Anoto con agradicimiento que Vm han remitido a los S[res] U. Zellweger le fer 7979.30 para los cuales les abono R 2174.20.
Sirvense Vms mandar los 30 barilos goma despues de su llegada a A.F. Matveieff de

mosca. Abono a Vm R 16.16 pagados
al Sr G. Chamot
Estoy muy agradecido – obligado a Vms para
sus ↓muy↓ interesantes informes.
A mi vuelta a Paris iré con mucho gusto
a Suecia para deliberar con el S[r] hijo
y hermano de Vm acerca de las propiedades
que Vms proponen. Sirvanse Vm
pues decirme con vuelta del correo la
direccion del Sr Sterky en Suecia. Si hay
de esperar una renta cierta de 4 o 5% entonces
compraré con mucho gusto; pero si no hay
ventajas ciertas no compraré, porque nada
de mas fastidioso que de trabajar de
balde.
San Ildefonso 29/10 Setiembre. Salimos ayer
a las 11 por la noche de madrid y llega-
mos aqui hoy por la mañana a las 7.
Teniamos buenos asientos de berlina por
los cuales sin embargo debiamos pagar 8$ ca-
da uno. en vez de encontrar aqui –
como esperaba – a un sitio real encontré
a una miseria. La casa real asemeja antes
a una casa ordinaria de dueño de aldea que a un
palacio real de verano. en el suelo – cuarto bajo
al menos todo esta sucia y parece en deca-
dencia. Se dice que los cuartos superiores
estan arreglados con elegancia pero no acerté
verlos. No podiamos conseguir el permiso de ver
[41]
los jardines, estando ellos hoy cerrados por motivo
del almuerzo que da la reina para el que estan
convidados mas de 300 personas. Fuimos pues
por un omnibus, a razon de ½ $ cada uno, a Segobia
que es una de las mas antiguas ciudades de España.
Alli esta el celebre acueducto romano de mas
de 150 piés de altura que consiste en grandes
arcos construidos los unos sobre los otros y la
gran maravilla es que las piedras estan unidas
sin cal ni cemento. Es incredibible como los
arcos asi hechos han podido sostener un fardel
– una carga tan pedosa [pesada] durante tantos siglos.
Ademas del acueducto merece atencion la
catedral con bóvedas de gran merito. Las balus-
tradas que circondan la Sagrestia, el coro y el gran

altar son de hierro y magnificamente adornados. En el coro hay gran numero de libros santas escritas con la pluma en la edad media. El pulpito es de marmol y muy bien esculptado.

Nos fuimos despues al Elkasar o palacio que forma una pequeña fortaleza montada de muchos cañones. El castillo es muy antiguo pero recien restaurado. Estan en el salas muy bonitas con cielos rasos a la moresca y endorados o de diferentes colores. Hay en las salas todo genero de modelos y patrones de arquitectura. Ofrece de los balcones una vista superbia sobre el valle. Fuimos despues tomar el chocolate en un cafe, donde todo estaba muy sucio. Volvimos a la una por el mismo omnibus en el que teniamos asientos de fuera espuestos a todo el poder del sol ardiente. A nuestra vuelta aqui presentóse nueva dificuldad, porque no sepiamos – sabiamos como salir, porque no hay ni omnibus ni diligencia que va al Escorial y para un coche por allí 7 leguas me pidieron 30$, lo que me obstiné no pagar. Me concerté pues con una mujer para que por 4$ me diese un buen caballo de silla y que su marido me accompañase a pié llevando mi saco de noche.

[42]

O que por 6$ mas me diese 2 caballos si mi compañero consentiera ir conmigo. Imposible describir cuanto me disgusta este viaje. Estamos aqui como vendidos y debemos permitir a cada uno robarnos cuanto le gusta. Volviendo hoy de Segobia en el omnibus tuvimos atras de nosotros algunos oficiales.

<u>El Escorial Domingo el 30/11 Setiembre 1859</u>

Ayer me acostó a las 6 de la noche porque estaba muy cansado y solamente me desperté esta mañana a las 6. Fuimos a las 7 ¼ a los jardines reales que en magnificenza mucho esceden aun los celebres jardines de Versalles. Al rededor de los jardines de flores los cedros estan cortados en forma quadrangular o redonda con un globo arriba. en medio de las flores estaba sentada sobre un peñasco la diosa de la independencia circondada de muchas otras diosas y dioses que

tenian ramos en las manos en cuyos bocas habia
cañitos o tubos por los cuales el agua corrió.
Habia una cantidad inmensa de semejantes
surtidores en todas partes en jardines
flores y en medio de las encrucijadas de las
veredas y nada ha de ser mas hermoso que el
paseo entre tantas maravillas cuando todos
los surtidores corren. Habiendome alejado de mi
compañero me aparté del camino y no sin
pena alcanzé el gran estanque donde habia
ayer el gran almuerzo que dió la reina. Todavia
habia alli la … y floreros o vasos con guir-
naldas y macetas. Los jardineros que habia alli
me esplicaron que habia mas de 600 convidados
y asi un numero demasiado grande para asentarse
a la mesa todos. Pues con excepcion de la reina
nadie se asento y todos almorzaran
en pié. El vino de Champaña colaba como agua.
Por estar rodeado de montañas y lleno
de encinas y fresnos este sitio es el mas fresco
de todos los sitios reales y por eso es justamente
preferido por la Corte. Salimos a las 9 ½ o
a las 10 and un coche que se asomejaba mucho
[43]
a aquellos coches de la edad media que vemos sobre
pinturas antiguas. Subiamos la montaña con-
tinuadamente durante 4 horas al paso y despues
bajamos con priesa y llegamos al Escorial a las
5 ½ de la noche. El cochero con quien habiamos
hecho la contrata que reciba 14$ si restasemos
solos o 13$ si tomase qualquier otra persona
no se quedó satisfecho de los 13$ que le entreja-
mos y pidio 14$, diciendo que solamente habia
tomado una persona and el pescante que le
pertenecia. No quería aceptar el pagamento
y fue quejarse al alcalde, el cual mando
por nosotros un alguacil con orden que nos
presentasemos al instante. El juez decidió
que pagasemos la mitad de la diferencia y la
mitad de la cita; lo que hicimos sin
contradigion.

El Escorial Lunes 31/12 Setiembre 1859.

Esta mañana salimos a las 8 teniendo
por guia un viejo del telor – oficio – de la pro-
fesion que ya habia estado ciego desde

43 años y que no obstante nos condujo en todas partes y nos hizo una descripcion minuciosa. Se diria que el ciego vé mejor que dos con ojos sanos._El edificio maravilloso que atrae los curiosos de todas partes forma un gran paralelógramo, cuya fachata es el palacio, los dos edificios laterales el convento de monjes y el en el fondo la iglesia. Hay en este gran edificio 10032 ventanas y puertas, 16 patios, 9 torres y 76 fuentes. Vimos luego el palacio cuyos cuartos se distinguen por la inmensa riqueza de los tapices bordados con la mano y representantes cuadros hermosisimos. El ajuar o los muebles y a frescos de las bóvedas y cielos rasos, los entablados que consisten de madera a modo de taracea – todo muestra magnificenza real. Cuatro cuartos que componen el gabinete y tocador de la reina se distinguen particolarmente por su inmense riqueza, allí cada cerradura es una obra maestra y cada puerta vale una fortuna. Vimos y recorrimos despues la
[44]
toda la iglesia con sus accesorios. Tambien este edificio presente maravillas de arquitectura. Por ejemplo hay un vestibulo cuya boveda es tan plano y llano que el empedrado. Los pulpitos o catedros consisten en marmol blanco y amarillo. Hay una infinidad de cuadros que son obras maestros. Hay cuatro organos. Sobre el gran altar hay estatuas de bronzo de los apóstoles. Solamente confieso que las paredes ↓desnudas↓ y columnas de granito pardo – canoso ofrecen una vista lánguida. Lo mas notable que vi allí es el panteon de los reyes, que consiste en una sala ronda de granito, cuyos muros estan rodeados de nichos en los cuales hay ataudes de granito, que parecen esculptados en los mismos muros. Cuando muere un rey y sus exéquias se hacen entonces remueven la cobertura y introducen el cadaver. Es un sentimiento singolar de verse rodeado de las cenizas de tantos reyes a quienes el mundo era demasiado estrecho y que yacen ahora silenciosos en sus angostos ataudes. Cuantas misas hay que hacer para proporcionar el reposo a sus almas turbulentos

y revoltosos! Como hoy es el aniversario de la muerte de Felipe II habia espuesta en la iglesia un califado cubierto de un lienzo negro con una corona. Volvimos a madrid por la diligencia Tuvimos asientos de banqueta de atras y fuimos aporreados por los golpes que el coche pedoso daba a cada instante sobre el camino desigual. Despues de mi llegada entré en un estanco de tabacos para comprar puros y puse mi baston sobre el contador, donde de repente disapareció. Me sentia muy enfadado – enfadoso, porque queria mucho a aquel baston.

<u>Cordova Viernes 4/16 Setiembre 1859.</u> Sali de madrid el miercoles 2/14 corriente por la diligencia del medio dia. Tuve un asiento de banqueta y podia ver bien aunque tenia mucho que padecer del polvo y de la incomodidad con que se bajaba y subia. La diligencia

[45]

fúe puesto sobre un carro del ferro carril. Llegamos a las 11 de la mañana a Tembleque de donde continuamos el viaje en Diligencia. Fuimos con gran priesa para adelantarnos – pasar adelante de la diligencia la madrileña, que salió al mismo tiempo con nosotros. La envidia que causa esta competencia es muy buena. Los caballos que estaban siempre prontos fueron unidos al coche en un golpe de ojo. Atravesamos la mancha que es desprovista – desnuda de arboles consiste en un país muy llano y poco poblado. Dos y tres leguas se puede ir sin encontrar a un pueblo. Los pocos pueblos y pocas ciudades que hay no caresen ni aseo ni lindas mujeres. La romance de D^{on} Quichote presta a todo aquel pais un hechizo indescribible. Durante la noche del Miercoles al Jueves atravesamos la Sierra Morena y a pesar de que era noche podiamos distinguir a la luz de la luna una multitud de hermosas paisajes. Atrevesamos Madridejas, Valdepeñas, Baylen, Andujar y etc. Aunque se vee bien de los asientos que teniamos, ellos son sin embargo peligrosos para tales – aquellos que no pueden trepar o sostener el peso del cuerpo enganchandose a los remos de cuero. Ademas el banco de madera esta solamente cubierto por una almohada de

cuero de modo que uno ha de sufrir mucho dolor and el trasero. Llegamos ayer a las 9 a Cordova, ciudad bastante limpia pero con un empedrado muy malo, sobre el cual no se puede ir and coche sino al paso. Las calles con estrechas y angostas para proteger contra los rayes ardientes del sol. El arreglo interior de las casas asemeja mucho el de las casas de Pompeo. en las casas de la clase mas que media no hay ventanas sobre la calle. Solamente hay una gran puerta adornada de una especie de clavos con cabezas grandes. Despues sigue un pequeño hermoso jardin con cipreses criados asi que forman figuras fantasticas. Alrededor del jardin esta la casa apoyada sobre columnas y formando asi grandes y anchas corredores, en los cuales hay siempre una deliziosa frescura. En las casa de los ricos los patios estan cubiertos
[46]
con una tela lienzo. en los cuartos hay ventanas a ambos lados de modo que se puede siempre tenerlos frescos por las cortinas y volantes. Chinches y pulgas no carecen.Y desde la Sierra morena donde empieza la Andalucia el terreno esta mucho mas feraz y and todas partes se puede cultivar centeno, seigle cebola, avena y trigo, pero el terreno conviene particularmente para las oliveras (olivos) de los cuales se ve todo cubierto por cuanto alcanza la vista. Las aceitunas se recojen and Enero y a pesar de que la cosecha no ha acertado en este año se puede decir que no hay planta que de en general una ventaja mas segura. Nos paramos – detuvimos en la fonda de Rizzi, que es la unica en la ciudad A pesar de los muchos chinches y pulgas dormi bien y fui esta mañana al baño en el guadalquivir. Despues del almuerzo me fui a la celebre catedral cuyas bovedas reposan sobre mas de 800 columnas provenientes del antiguo templo pagano de Jano, que estaba en el mismo lugar. El el construir esta mesquita los arabes querian construir una que excediese en grandarie y belleza aun la de meca. Han conservado en ella todavia algunas capillas y entre ellas aquella and que habia

نقرقلا Allí hay todavia muchas inscripciones arabes. La piedra esta un poco ahondada – ahuecada – excarvada sobre la cual los arabes frotaban con el pié. Hay 19 altares y el gran coro, cuyas maderas son esculptadas con sumo gusto – sumo talento. Representan los santos de Cordova y la historia de la biblia. Hay un cadelabro de plata pura que pesa 26½ arrobas. en las columnas de porfido hay incisiones que representan la santa crucifixion y que se dice ser hechos por presos-prisioneros con las uñas. Y esto no es estupendo – asombroso si consideramos que por las besos, que son mas suaves que las uñas, se ahunda el porfido. La una de las incisiones que es mejor hecha que la otra, esta protegida de pequeñas barras de hierro y al lado esta
[47]
un bajo relievo que representa el preso. Lo maravilloso de las incisiones es que los presos las hicieron por detras de las espaldas siendo encadenados por las manos atras del torso a grapones – rampiones en las columnas. en el patio de la mesquita hay magnificas alamedas de naranjos. Pregunté el camarero que tiene al empeine un genero de hinchazon, que tiene, me dijo que las tripas habian salido y pendian ahora fuera.

<u>Sevilla Sabado el 5/17 Setiembre 1859.</u>

Salimos de Cordova ayer a las 6 ¼ de la noche por el ferro carril, ibamos muy lentamente ademas teniamos en todas partes grandes detenciones, como por ejemplo ¾ hora a Posada, de modo que no alcanzamos Seville antes de media noche. Pero estabamos suficientemente remunerados por las conversaciones que teniamos en el camino con las mujeres y particularmente en Posada. Estaba en el mismo coche con nosotros el carpintero de obra prima Juan Nepomuzeno Causino plaza S. Fancisco N° 7, que me habló de una hacienda que se vende una legua de aqui y que, segun dijo es buen propiedad como utilidad y recreo. Prometio mandarme hoy por la mañana su hijo a fin que me acompañase. Este vinió pero se observé inmediatamente grandes

diferencias entre lo que me habia contado el padre. Prometió procurar un permiso para ver la hacienda mañana por la mañana y irémos entonces juntos a caballo. Fuimos a la Catedral, que es una de las mas hermosas que he visto. Tiene bóvedas muy altas y 91 vidrieras pintadas. La sagristia, el coro y el altar mayor son obras maestro de esculptura de madera. Se distingue la crucifixion de Cristo esculptada tambien en madera. Todo está rodeado de vergas de bronze o de hierro artificialmente tornado.
En la capilla real se veneran los cuerpos de Fernando y otros reyes. Una capilla lateral se distingue por su riqueza en plata. Las vergas, el altar y los grandes candelabros consisten en ese metal. Entre muchos quadros hermosos hay uno, el cual los ingleses prometieron
[48]
cubrir de onzas de oro, si la iglesia qiera venderlo a este precio. Muchos adornos en stuco son imitaciones del estilo gotico. La iglesia tiene 431 pies de largo, 139 de altura, 171 de alto en el cimborio e 315 de ancho. Fuimos despues al museo de pinturas, en el cual no hay sino dos salas, de las cuales la una contiene esclusivamente pinturas de Murillo. Despues vimos los jardines del alcázar llenos de hermosas alamedas de naranjos, mirtos etc. en todas partes hay surtidores que el jardinero hace correr por una presion con la mano sobre el cañuto – tubo para regar las plantas. Vimos despues el Alcázar, de que tanto la fachada que los cuartos interiores merecen un viaje al rededor del mundo. Las paredes de la fachada, las de los cuartos y los cielos rasos estan cubiertas de bajorelieves moriscos en estuco entremezclados de versos del القصر- todo pintado con vivisimos colores, que estan entremezclados de doraduras. Indiscrebible [indescriptible] hermoso es el patio. Alli me mostraran la salió [sala] en que la duquesa de monpensier Orleans hermana de la reina parió and 1848. Ciertamente habia escogido para el parto ese notable sitio para dar todavia mas realze al ilustre parto, probablemente en la esperanza que pariese un hijo y que este subiera sobre el trono de España. Llovia y volví pues en casa para anotar las cosas

notables que habia visto. Nada en verdad de mas hermoso que las pequeñas, estrechas y sinuosas calles por la mayor parte empedradas o embaldosadas con piedras platas y llanas mantenidas con gran aseo y limpieza; nada de mas agradable que las casas construidas de ladrillos y cubiertas de cal blanco para mejor absorver los rayos del sol ardiente. Las casas no tienen ordinariamente mas de dos pisos. en el suelo o cuarto bajo hay dos ventanas protejidas al esterior por vergas de hierro pintadas de color verde y al interior por ventallas y cortinas. En medio de la casa hay una gran puerta de madera pintada morena, o de vergas pintadas verdes que abre sobre un corredor ancho que conduce al patio embaldosado de tablillas de marmol. En medio del patio hay una fuente. El patio esta cubierto de toldos, que se puede estrechar o estender facilmente por medio de una cuerda.
[49]
Al rededor del patio hay hermosas galerias que sirven de habitacion durante los calores del verano y estan amuebladas con sillas, canapés, quadros, jaulas de pajaros, macetas de flores, arbustos esoticos, faroles – todo colocado con mucho gusto. Las galerias estan apoyadas por columnas y a cada paso el estranjero se detiene delante de las vergas o puertas abiertas para admirar los hermosos patios y disfrutar de la vista. en el segundo piso no hay ventanas sino puertas vidrieradas que abren sobre pequeños balcones. Hoy por la mañana paseandome sobre el mercado de las frutas encontré a una mujer que iba con una joven de 14 bastante linda pero de un esterior triste y melancolico que me intereso. Me acerqué a ellas y propuse a la vieja darme la chiquita a fin de que viviese con ella en Paris. Ella respondió que para eso era menester que nos conociesemos un poco mejor. Ella me dió la direccion de su casa y me fui a ella la noche. Ya hallé las dos en la calle y ya me pareció que aconsentian y empezaba lusingarme que no harian obstaculo. Estaba pues mucho admirado – pasmado – asombrado cuando oí que ni la hija queria ir ni la madre queria separarse

de ella por miedo que los vecinos no criticasen su conducta de abandonar a su hija tan joven. Cuando vi que todas mis persuaciones no alcanzaron mi intento les propuse ir con ellas al baile. Ellas querian ir solas y no me opuse. Me enternecé veyendo asentada sobre el banco de los espectadores mi querida con una cara enfadada, enojada y triste. De balde me esforzé demostrar a ella y a la madre que terrible tiempo venidero aguardaria la pobrecita si reechaze – rechazase mis buenas proposiciones que no tenian otro intento que la dicha de la chiqua. Pero todo era in vano. De balde me esforzé esplicarle que seria seducida por un miserable y abandonada y que se lastimaria y se arrepentiria toda su vida. Eran bailes miserables y sali pronto siguiendo la vieja con su hija que traté con sorbetes aprovechandome todavia de esa ocasion para demonstrarles cuan buena fuese la vida and Paris. Temiendo que mi pasion para la hija no tomase
[50]
echase raices demasiado profundas queria separarme de ella ahora cuando el amor acabó solamente de hacerse sentir y me despedí pues de ellas para siempre.

Sevilla Domingo el 6/18 Setiembre. Salí este mañana con el corredor con direccion a Bermuza para ver la hacienda de Venen. El hombre habia jactado mucho con su destreza en el montar a caballo y estaba pues muy pasmado de ver que a penas podia trotar. El daba la culpa a la silla diciendo que no estaba acostumbrado a emplear otras sillas que sillas españolas. Este paseo pertenece a los mas agradables que he nunca hecho en mi vida. La vista de la montaña sobre la ciudad es hermosisima. Tambien la hacienda me gusta; hay mas de 4000 olivos o oliveras y dos grandes campos plantados con viñas. De ambos productos la cosecha de este año no ha acertado bien y las oliveras ofrecen a penas la tercera parte de la cantidad de aceitunas que habia hace 2 años. El año pasado aun no habia mas de 35 arrobas de aceite

pero el uno y en el otro creo que se pudiera
calcular sobre una renta anual de las oli-
veras y de las viñas juntas de almenos 1000$
lo que no seria mal si la suma fuese mas
grande. Si compro esa hacienda lo hago prin-
cipalmente para hacer el amor a las guapas
andaluzas, pero tengo miedo que este recreo
me fastidiara pronto, porque esta aqui cir-
condado de gente estupida y nó hay ni caza
ni pesca. La casa del amo no vale nada
porque tiene solamente tres cuartos
pequeños y debiera antes todo otro
añadir todavia un piso. Todas las maqui-
nas para la fabricacion del aceite y del
vino estan en buen estado y parece que los
espesos muros desafian los siglos. El encar-
gado me parece hombre muy honrado y
no pudiera desearme un mejor. Jardin no
[51]
hay sino algunos naranjos en el patio.
Fuimos despues a otra hacienda cerca de
la ciudad que consistia en poco mas o menos de
800 naranjos, peros y et, entre los cuales habian
sembrado maiz, legumbres, hortalizas, centena
etc. Hay dos casas de dos pisos en buen estado.
Rogué el corredor informarse acerca del precio
de esa hacienda.Volviendo despues a la ciu-
dad pasamos el magnifico palacio de la
Infante con grandes jardines. Dí despues
buen almuerzo a mis compañeros y sin
cambiar mis vestidos me fui a la corrida de
toros, donde tenia mal sitio y me desmayé
del calor. Me pareció que aqui los toros son
mas silvestres y salvajes y que tanto los
torreros que los picadores trabajan con
mas destreza y mas de priesa que en Madrid.
La alegria era tan grande que cerca de mi
varias veces personas empezaron a pelear
entre ellos furons conducidas fuera por la
policia. Salí antes del fin y me fui despues
de la comida al teatro San Fernando, donde
representaban: "el Cura" cancion o romance
en versos. El asiento en el patio no costaba
mas que 3 ^{rs}. Me gustó mucho la representacion
y tambien el teatro el cual era bien adornado.

Todos los sitios en el patio eran sillones o sitiales doblados con peles y cubiertos con cuero.

Sevilla Lunes el 7/19 Setiembre. Cansado por los muchos gozos que disfruté ayer me desperté hoy tarde, almorzé y me fui a la Catedral en la cual me hice indicar por un fraile emancipado todas las cosas notables, entre las cuales llamó, tocó – picó mi atencion el sepulcro del hijo de Cristobal Colon. Una inscripcion sobre la piedra dice que habia dedicado toda su vida a las leteras y uniendo en una obra de 4 tomos todas las ciencias entonce conocidas la legó junta con su biblioteca a la ciudad de Sevilla. Morio en 1538 en edad de 50 años. Allí es hecha tambien mencion de los dos buques de Cr. Colon, cuyos dibujos hay allí. Son unas
[52]
barcas pequeñas de dos palos, que tenian juntos 90 hombres de tripulacion. Estaban construidas para ser remados en tiempos de calma como lo indican los muchos marineros con los remos en los manos. Hay alli grabado en la piedra todas las armas de Cr Colon como gran almirante vicerey. Vi despues las alhajas de la catedral consistentes en plata y trages recamados – bordados de sacerdotes y subí en la torre. en vez de escalera hay un camino de caracol ancho y no escarpado el cual se pudiera montar a caballo si la puerta de la entrada no fuese demasiado bajo. De la galeria de la torre, donde pendono las campanas, se ofrece una vista superbia sobre Sevilla con sus millares de casas blanqueadas con cal para diminuir el calor del sol. De allí me fui a la gran fabrica de tabacos Un conserge o custado me condujo luego en todos los cuartos del piso bajo – cuarto bajo – suelo en el una hay herrería o forja, en otra hay fabrica de cajas de hojas de lata en las que empaquetan el tabaco de nariz, en otro hay almacenes de tabaco en rollos, en otro preparan el tabaco con azucar y otros ingredientes y en otros lo pican en maquinas movidas por mulas. Subimos despues al segundo piso donde hay inmensos salas llenas de mesas a las cuales tra-

bajan las mujeres. Las unas quitan los tallos
de los hojas, otras envuelven o tornan
los cigarros, otras hacen de papel sellado por
medio de una forma sacos de papel para los cigarros.
A cada mesa hacen solamente un trabajo y
por consiguiente lo hacen con gran velocidad y
ligereza. No puedo negar que el espectaculo de
tantas mujeres era el mas interesante que dis-
fruté en mi vida. Casi todas son perdidas
pero entre ellas hay muy guapas. Juzgando por
los ojos hundidos y al parecer venereo de las
mujeres reconocí que cási todas estaban en-
fermas, lo que es natural porque no pueden
vestirse de los 3^{rs} diarios que ganan y deben
luscarse en las calles un suplimento de ga-
nancia ↓o lucro↓. A pesar de su miserables vidas y
de sus esperanzas y prospectos todavia mas mis-
erables para el tiempo venidero y a pesar
de las infermedades con las que estaban
[53]
afligidas, todas parecian muy alegres y estallan
de risa cuando vieron que las miré
regendome [?] de las unas y compadeciendome de
las otras. Varias veces preguntaba las que eran
mas guapas si querian irse conmigo a Paris
y las hallaba – encontraba siempre prontas. La maestra
(que me acompaño para impedir desórdenes
o escenos de la parte de las muchachas) tose mucho
tiene el pecho hundido y es incurable porque
el olor fuerte, que siempre ha de dolor, la mata
y ella moriró pronto de la tisica. Me fui
despues a la lonja, Donde habia ayer grande
comida and honor de la procesion que vinó
por la primera vez sobre el ferro-carril de
Jerez. Las mesas estan todavia and pié. La
lonja esta construida como todas las demas
casas de esta quiero decir con pátio en el medio
rodeado – circondado de galerias apoyadas sobre
columnas. Todo el segundo piso esta lleno de
armarios que contienen los archivos de las
provincias españolas en america desde
1500 hasta ahora y todos los avisos y infor-
mes dados por los grandes descubridores y gene-
rales, todos los tratados hechos etc. Cada
suerte de papeles esta puesta seperadamente

todo esta envuelto en paquetes o lios, sobre los cuales esta escrito de que asuntos trata y a que epoca pertenecen. Del segundo piso al tercero conduce una escalera admirable que consiste de largas piedras juntados o unidos sin el menor apoyo y sin que haya en toda la escalera la menor curva o boveda que pudiera sostener el peso inmenso. La unica cosa que segun mi parecer pudiera dar algun sosten es que el lado abajo de las piedras esta cortado o tajado de modo que forma una superficia oblicua – torcida semejante a la aspa de un molino. Me fui despues al banco, donde vi puesto un rótulo que el discuento para letras de cambio es solamente 6%.

Cadiz 8/20 Setiembre. Por fin he alcanzado esta linda ciudad, a la cual nunca hé podido pensar sin gozo porque todos que estaban aqui me han hecho de ella elogios – alabanzas

[54]

hasta el cielo. Salí de Sevilla esta mañana a las 8. La vista sobre Sevilla con su hermosa giralda era muy hermosa en verdad cuando partimos en El Guadalquivir se percibe la marea hasta Cordova. Sobre ambos lados del rio habia hermosas haciendas. El rio no es mas ancho a Sevilla que la Sena a Paris, pero se ensancha poco a poco hacia el embocadero. A poco mas o menos de medio camino de Sevilla al mar el terreno se aplana y es espuesto a las inundaciones del rio. Alli por cuanto alcanze la vista no se ve nada que lagunas y prados y turbas o rebaños innumerables de pajaros y particularmente de patos. Teniamos sobre el vapor buena alegre compañia y en continua conversacion el tiempo se pasó de priesa. A San Lucar bajó la mayor (2) parte (1) de los pasajeros, que, teniendo el movimiento del buque y el mareo, preferió hacer el viaje por tierra aunque con un rodeo o rumbo grande. Cadiz con sus innumerables buques con todos las pabellones del globo y sus magnificas limpias casas, ofrece una vista hermosisima a el que viene por el mar, tanto mas cuanto todo el espacio que la naturaleza tiene asignado

a la ciudad esta ocupado de edificios que parecen pues estar construidos sobre el agua. Me paré en la fonda de Blanco, que no es sino de 3 ra clase and comperacion con la fonda de Paris a Sevilla. Sin embargo mi ventana tiene la vista sobre el mar. Desde las 6 hasta las 9 me paseaba (yo daba una vuelta) buscando aventuras amorosas. 1/1/2 horas gasté sobre la hermosa alameda donde todo el hermoso mundo de Cadiz se paseaba bajo el tocar de la musica. Las señoritas parecian hacer ostentacion a porfia de sus hermosas caras, grandes ojos, que brillaban como otros tantos soles y la sensillez de sus trages españoles con sus largos velos daban todavia mas realze a su hermosura. en presencia de
[55]
tanto lujo de la naturaleza ni podia pensar a pegarme – agorrarme a una de las muchas mujeres infelices que encontraba a cada paso y que estaban prontas a regatear conmigo por el precio de su amor. Dos veces abordé señoritas decentes pero no me respondieron y pues no me atreví mas probar a entablar una conversacion con ellas. Cadiz Miercoles el 21 Setiembre. Estaba tan cansado de las grandes vueltas que daba ayer que me desperté esta mañana solamente a las 7 ½ y me levanté a las 8 horas. Fui al establecimiento de los baños de mar, donde disfruté mi baño mucho. El agua del mar contiene tanto sal que facilita mucho el nadar. Me enjugué – sequé despues con una toalla, almorzé huevos fritos con tomatos, pescado asado y chocolate y me puse despues en camino para ver la ciudad. Di luego una vuelta sobre los baluartes al rededor de la ciudad, atravesé la hermosa alameda donde habia ayer oido tocar la musica y visto las guapas; la vista del mar que reaviva en mi tantas recordaciones me llena siempre de gozo. en todas partes estan puestas cañones para defender la ciudad, que fui saqueada una vez de los franceses y despues sitiada en vano varias veces. Donde se habia ahondado – ahuecado agujeros para reparar las fortificaciones o para plantar arboles ví que la tierra esta llena de destrozos de ollas, vajillas de barro, ladrillos, cal etc y pues es

dificil cultivar aqui jardines. Mas adelante hay dos castillos, de los cuales el uno esta contruido sobre peñas; los fundamentos o substenciones se dice que son hechos por los fenicios. Recorrí despues la ciudad en todas las direcciones y encontré en todas partes mujeres guapisimas vestidas de sarga negra con un gran pañuelo de seda negra que pende hasta el trasero y un velo al rededor de la cabeza. Sus cabellos negros o rubios que alcanzarian (la tierra) – el suelo si fuesen desatados son ligados detras, en inmensas soguillas. El tamaño de sus piés es muy chiquito. Por estrechas que sean las calles hay sobre ambos lados ánditos – andenes de piedras platas. El mar refresca – resfria aqui el aire. Las casas son
[56]
muy hermosas en partes con techos platos a partes con techos oblicuos con tejas. Aunque Cadiz me parece un paraiso no podria vivir aqui sin ocupaciones fijas que no podria proporcionarme sin pagar caro para mi aprendizaje. Sobre los muros que circondan la citadela son asentados siempre centenares de hombres que pescan a la caña, con hilitos y con redes.
Cadiz 22 Setiembre 1859 (Jueves). Salí esta mañana a las 8 horas 10 m. con direccion a Jerez. Todavia no estando terminado el ferro carril hasta el muelle habiamos de atravesar la bahia en un vapor hasta el embarcadero del ferro carril, que esta en el sitio donde los franceses en 1810 establecieron sus baterias y bombardaron Cadiz. La distancia es muy grande y parece incredible que las bombas hayan podido alcanzar la ciudad. Atravesamos un terreno – termino llano en el que eran ahundado – ahuecado gran numero de zanjas o fossos para secar el termino. Despues vimos a ambos lados nada que viñas que producen el deleitoso vino de Jerez. El terreno era arenoso, no obstante las vides parecian crecer bien. La ciudad de Jerez no ofrece absolutamente nada de interesante sino las bodegas de vino. Vi solamente las 5 bodegas del consul francés Domecq que son edificios muy grandes llenos de barricas

o botas de vino. El encargado sacó con una cuchara un poco de vino de varias barricas que me dió a beber y entre otras me dió a probar vino 60 años en edad, que estaba espeso como jarab o almibar. Ví tambien como hacen las barricas, como ponen sobre ellos los anillos de hierro, como las queman en el interior etc. Visité tambien el Casino y volví a las dos a Cadiz. Por la noche habia música sobre la plaza con alamedas y se veia allí toda la flor de las hermosuras femininas. Me acerqué de una y no sin dificuldad acerté a entablar una conversacion que se hizo mas intima cuando empezé a hablar de trages de mujeres. Como ví la inposibilidad
[57]
de obtener favores particulares me alejé – aparté pero no sin haber tomado la direccion de la niña que era Eliza, calle larga N° 50.

Algeciras 23 Setiembre (Viernes). Salimos este mañana a las 6 ½ sobre el vapor Bernardo con direccion a gibraltar. No sin pesar – arrepentimiento me despedí de Cadiz y trepé en el palo para ver la ciudad cuanto fuese posible. Una voz me dijo que no lo volveria a ver. El vapor, que tiene solamente la fuerza de 25 caballos, marchaba muy lentamente. Cadiz dispareció y ya vimos la costa montañosa de Africa. El viento que se volvió contrario y sopló con furor por el estrecho ralentó todavia mas nuestro viaje. Dichosamente la corriente nos ayudó de otro modo no hubieramos podido adelantar un paso. Hicimos escala – descansamos un ratito en Tarifa y llegamos a las 5½ de la noche a Algeciras en cuyo puerto habia varios vapores españoles que habian llevado ejercito y municiones para la guerra en maruecos.

Gibraltar Sabado 24 Setiembre. Llegué aqui esta mañana a las 9 y despues de haber almorzado yo me puse en camino para ver las formidables defensas de este nido de aguila. Por desgracia – desgraciadamente era menester un permiso, que el gobernador de la plaza no entrega sino a la peticion – demanda del consul y como no encontramos lo en casa y debia aguardar no alcanzé el permiso antes de las 11. Fui luego con mi guia un judio ver las galerias subterráneas ahondadas –

ahuecadas en el peñasco que domina el estrecho
istmo pantanoso -charcoso que junta a gibraltar con
la tierra firma, ahondadas en linea oblicua las unas
sobre las otras. Las aberturas a las cuales estan
puestas los cañones parecen muy grandes en las
galerias y en verdad ellos son bastante grandes para
dirigir con comodidad las mas grandes piezas, mien-
tras que por la gran altura de 500 a 700 pies en
que estan no parecen de abajo sino como agujeros de
pichones. Por eso es dificil digo imposible alcanzarlos.
De tiempo en tiempo se sale de las galerias sobre
playas libres bien surtidas por baterias al aire libre.
Por fin se baja en las galerias subterraneas un declive
de poco mas o menos de 100 pies y se llega a una
sala grande redonda llamada george Hall surtida
[58]
con poco mas o menos de 20 cañones de gran ca-
libre, que dominan el mar en la vecindad del
istmo. Subimos despues sobre la cima del peñon
donde hay una bateria inmensa con cañones de
gran calibre. Ha de ser enorme la fuerza de las
balas arrojadas de aquella altura de 1460 pies
y en el caso de un ataque preferiria hallarme
antes sobre el peñon que en los buques,
porque la fuerza de los proyectiles, que
aumenta por la casi perpendicular caida, ha de ser
tanta que las balas atraviesen las mas fuertes buques
de hierro. A bajo hay tambien tanto a aire libre
sobre el peñon que en el pueblo varias baterias
fuertisimas para proteger el istmo y pues por
esta parte (por este lado) la fortaleza es inexpugnable
sino por traicion. Subimos despues sobre el peñon
vecino cuya cima esta poco mas o menos
de 1250 sobre el nivel del mar. A pesar de que
hay un camino caracol o camino que va and direc-
cion oblicua y torcida es muy dificil subir y nos
cansamos mucho; sudamos como si estuviesemos estendidos
una hora en el baño de vapor. La vista es hermo-
sisima y abraza no solamente las montañas del
mediodia de España sino tambien la costa de
montañosa de maruecos y la inmensa superficie
del mar. Allí hay una pequeña bateria con 3 o
cuatro cañones de gran calibre y una casa en la
que viven los tres guardianes o custodios, que
anuncian siempre la entrada y salida de los

buques por el levantar de bolas y banderas. Miran siempre en todas las direcciones por el telescopio, el que es de gran portanza y apenas descubren en la lontananza un buque que ya advierten de el levantando una bola. Mientras estuve alli se acercaron a el puerto y entraron en el 8 grandes vapores ingleses de guerra, de los cuales cada uno tenia 76 cañones y aun la real tenia 141 cañones. La real fúe salutada por las baterias a tierra por 17 cañonazos y respondió al saludo por otras tantos. Venian aquellos vapores de malta. Los guar-
[59]
dianes eran muy afables particularmente cuando bebi una botella de cerveza por 15 . El uno de ellos vivia allí con su familia y de sus dos hijos el uno habia nacido allí, no tenia mas que 11 meses pero ya iba bien y hablaba sin mucha dificultad, lo que prueba que el clima es allí sanisimo. Hay allí una fuente delante de la casa. Si uno mira del peñon a bajo por la parte del mar parece que el peñon es accesible, pero no es sino una ilusion, porque hay 600 piés de arena, despues 400 de peñon escarpado y por fin 250 piés de peñon que se pudiera subir. Fuimos despues al tercero peñon, que es el mas escarpado de los tres. No hay en su cima otra casa que una torre antigua construida por los moros. Sobre este peñon por la parte del mar en frente de Marruecos hay una bateria muy fuerte montada por 74 cañones de gran calibre. Bajamos y dimos la vuelta al rededor de los peñones abajo, pero no se puede ir mas lejos que hasta la bateria llamada 4th Europe advance que esta casi al fin del peñon sobre cuya cima hay la torre moresca. Allí hay 3^{d} , 2^{d} y 1^{u} Europe advance y gran numero de baterias con otros nombres. Entre la 2da y 3^{en} Europe avanzada hay la casa de madera del gobernador guarecida por los peñones de los vientos del norte y oeste. Con escepcion de algunas de aquellas plantas, cuyas grandes hojas se emplea para hacer cuerdas, no hay vegetacion sobre aquel termino montañoso cubierto de muy poca tierra. Volvivos dando todavia una vez la vuelta de las baterias por el lado del mar y subimos despues sobre el gran peñasco que casi horizontalmente

se estiende 400 pies mas arriba que las baterias de abajo o 500 pies arriba del nivel del mar. Al rededor de este gran llano por encima hay cañones de gran calibre. Tambien hay allí un gran cuartel circundado de muros, con aberturas para los tiradores con escopetas y poco mas o menos de 35 aberturas para tirar con cañones. Bajamos por una escalera caracol y continuamos visitar las baterias que hay por la parte de la bahia. Aunque haya allí gran numero de baterias bien construidas con rincones y esquinas para hacer resbalar las bolas, las baterias no pueden ser muy fuertes
[60]
por razon de la gran estension de la costa y aun hay un gran espacio donde del todo no hay baterias y mi parece que allí se pudiera trasladar y desembarcar ejercito y tomar gibraltar por asalto, porque todas las baterias en las galerias y todas aquellas que defienden las otras partes no serian de ningun uso si g. fuese sitiado y embestido por esta parte. Ya han visto los ingleses la nececidad de tener alli defensas y ya estan contruyendo baterias allí mas arriba, pero hasta ahora no han terminado el trabajo. Hay en medio del pueblo el jardin o la huerta del gobernador, cada uno tiene acceso a este jardin, que es el paseo de predileccion de los habitantes por la noche y, como me dicen, hay allí siempre musica que toca. La ciudad esta bien construida y hay algunas casas muy hermosas. Hay algunas calles rectas y empedradas con piedras platas con pequeños ánditos o andanes por ambos lados; pero como la ciudad esta construida sobre el declivo de un peñon es natural que por la mayor parte las calles son escarpadas y torcidas y muchas de ellas mas arriba aun cubiertas con arena en vez de empedrado. Siendo gibraltar un puerto franco pensaba recibir allí todo casi de balde; pero me engané aun los precios de los cigarros no hay que poco diferencia contra los precios en España y zapatos y trages son todavia mas caros. Por la noche me fui al teatro, que es de madera y muy pequeño. Representaron una zarzuela con cantos. Un pasajero que habia ido con mi y que va hoy a malta, habia

tomado los billetes de segundo palco, donde
no hay sino putas. Hablamos mucho con esas
desdichosas, pero me fúi a las 10 porque
estaba cansado. La plaza costaba 5 .

En el mar entre Malaga y Gibraltar
Domingo 13/25 Setiembre. Salimos este
mañana a las 7 por el vapor
San Bernardo con direccion a malaga
y haciendo el rumbo de gibraltar vi
[61]
todavia una vez sus hermosas baterias.
Malaga Lunes 14/26 Setiembre. en España tienen
la mala costumbre a dar el pecho a los niños
hasta la edad de 2½ y aun 3 años. Pienso que esto
ha de ser muy malo, sin embargo como el gran
calor produce frequentemente inflamaciones en el
celebro en el tiempo cuando los niños hacen
dientes, puede ser que les dan el pecho tan largo tiempo para aliviar y facilitar el crecer de los dientes. Aqui no
hay absolutamente nada de notable, si no es la
gran cantidad de hermosa fruta con que las calles
y plazuelas estan sobrecarregadas y la
mejor uva se vende a 3 Cop la libra. Ayer
a mi llegada me apresuré ir al correo, donde
sobre las listas de cartas descubrí mi nombre.
Pidé la carta el secretario pidió que
me legitimase lo que hice sin plazo – dilacion.
Era una carta de mi agente de moscow, a la que
respodi hoy mismo. Por la noche me fui de nuevo al teatro y mi mala quizo que a pesar mio
tomase un billete para el segundo palco donde
habia de calentarme entre dos putas, de quines
la una parecia ser la ama y la otra la servidora.
Mucho me disgustaba esta sociedad. Cuando volví
del teatro me acosté, porque estaba cansado por no
haber bien dormido varias noches. Pero apenas
habia apoderado de mi el sueño – apenas dormi cuando me desperté por un dolor terrible
y cosquillas sobre todo el cuerpo. Encendí la
vela, miré el lienzo de la cama y ví con horror
que todo estaba lleno de chinches de todo tamaño
y calibre. Como no he tomado conmigo el polvo
o polvora contra insectas pusé juntas las sillas
y me acostó sobre ellas tomando el saco de noche
por cojin – almohada. Pero por causa de los rin-

cones de las sillas esa cama improvisada me pareció demasiado dura y me acosté pues sobre las dos mesas que junté. Pero, como se puede suponer dormi muy mal. Esta mañana amenazé tomar otra fonda si no me diesen otro cuarto sin chinches. Me dieron un otro pero ni sé si no hay en el chinches tambien.
[62]
Los cuartos estan aqui semejantes a los en Cairo, pero mucho mas sucios. Sobre la cama hay un hilito de zancudos, lo que es necesario porque hay abundancia de esos insectos, que muerden y fastidian mucho. Esta mañana me fui al baño de mar que es muy bueno. Me dieron una sabana para enjugarme. Despues escribí algunas cartas, comé y me fui al teatro.

Granada Jueves (el dia de San Miguel) el 29 Setiembre 1859. Sali de malaga a media noche del 27 al 28 corriente en el correo que es un miserable carro con 2 ruedas, con el mayoral, el postillon y un pasajero delante y 4 pasajeros detras en un espacio en que apenas caben 2. Era uno de estos ultimos infelices y no puedo describir quanto hé padecido del calor, del polvo y de la estecheza porque no habia lugar donde poner los pies. en fin la compañia era muy mal escogida y habia de contenerme para no ofender el uno o el otro en su estupido orgullo. Las primeras 3 horas de malaga subiamos continuadamente el declive bastante escarpado de una montaña y con pocas interrupciones teniamos durante todo este tiempo las luces de malaga en vista. Despues bajabamos y subiamos alternativamente hasta que por fin llegamos a la hermosisima llanura de granada. Pasamos el pueblo de Santa Fé, donde Fernando el Catolico venció al ultimo rey moro y conquistó a granada y media hora despues entramos a granada. Tenia la intencion de pararme a la fonda de la Victoria, pero al correo habia el amo de la fonda de la Minerva con un guia de estranjeros, que me dijeron, a

mi pregunta me dijeron que en la fonda habia un frances que segun su descripcion habia de ser el mismo Melchior Mocquai con quien habia viagado desde madrid hasta
[63]
Sevilla. Preferí pues alojarme en la minerva donde estoy bastante satisfecho. Despues de la comida me fui a una casa de putas donde me puse un inmenso cordon sobre el carajo y godé una joven bastante linda, ho hice solamente para poder decir que habia disfrutado una castellana. Me lavé pues y enjugé – sequé, volví a la fonda y me acosté. Esta mañana despues de haber tomado el chocolate sali para la alhambra situada un poco fuera de la ciudad sobre una colina grandiosa y magnifica es la entrada por el parque que esta adornado de hermosisimos fresnos y otros arboles. Subiendo la vereda a mano izquierda pasé la puerta llamada: de la justicia y entré en un largo corredor abierto, que me conducé a la plazuela de los arquives. Cerca de esa está una gran plaza llamada Alcazaba, que evidentemente estuvo lleno de edificios de la antiguedad. Subí despues la torre de la vela con escalera caracol, de cuya cima se ofrece un magnifico panorama sobre la ciudad y los contornos llenos de haciendas, de huertas y plantaciones y sobre la Sierra nevada cubierta de nieve perpetua. Bajé despues y me fui al palacio arabe, a cuyo lado hay el palacio de Carlos V que ha nunca estado terminado. El esterior soberbio annuncia un suntuoso interior, pero subiendo allí por la escalera se viene a un inmenso patio redondo con galerias al rededor apoyados abajo de 32 y arriba de un igual numero de columnas de marmol. Dentro hay nada que ver porque niguna sala esta terminada. Ya antes de subir la torre de la vela habia encontrado a un guia, con quien habia contratado de conducirme – guiarme mediante 5 fr hasta las 5 de la tarde. Entré en el palacio arabe por una escalera de 27 escalonas cada uno de 3 varas de ancho y todas de jaspe de una sola pieza. Entré en el patio de los Arrayanes o mirtos, en el que hay un estanque con pescados rodeado con arrayanes; a los

dos lados de mano izquirda y de mano derecha
hay 8 columnas de marmol blanco. Entré despues en el patio de los leones, llamado asi
[64]
por una fuente o un surtidor con 12 leones que hay en el medio. Cada uno de los 4 lados de este patio forman una hermosa galeria apoyada con 32 columnas pequeñas de marmol blanco. A dos lados hay una suerte de pabellones o templetes con surtidores en el medio. Cada pabellon reposa sobre 22 columnas. A mano derecha esta la sala llamada de los Abeneorrages, en que segun la tradicion, fueron degollados 36 abencerrages; – en esta sala hay una fuente con larga mancha morena que se dice estar producida por la sangre de los degollados. Al otro lado hay 3 cuartos cuyos cielos rasos son adornados de pinturas arabes, en las que figuran hembras y varones, lo que me pereció sumamente singular porque el profeta mohamed ha severamente prohibido de pintar criaturas humanas para impedir el paganismo. en el una de los dichos cuartos llamado: "sala del tribunal esta pintado un castillo con doncellas que miran fuera; a un lado un caballero o ginete en armadura traspasa un otro con su lanza; al otro lado un ginete mata una fiera que tiene una doncella robada en sus brazos; en el otro cuarto hay cielo raso que represente tambien un castillo con doncellas; sobre el empedrado hay un gran jarro vaso de porcelana; en el tercero cuarto el cielo raso representa una caceria. Al otro lado hay la sala llamada de las dos hermanas por dos piedras de marmol iguales en grandarie [grandeza]; a ambos lados hay pequeños cuartos; en las paredes de la entrada hay nichos en los que se deponia el calzado. Fuimos despues a la sala del Andarage a la cual se junta el jardin del Andarage con hermosas fuentes y naranjos. Allí hay un patio con unas rejas que dicen ser la prision en que fue encerrada la reina. Subimos despues al tocador de la reina restaurado por Fernando y Isabel con muchas pinturas arruinados por los nombres arañados – escritos por los viajeros.

En un rincon hay una pieza de hierro con
muchas agujedos que se dice haber servido para-
[65]
hacer entrar en el cuarto los perfumes y fragancias de abajo. De los balcones de esta sala se ofrece una vista deliciosisima. Fuimos despues al baño del rey, que es un bacin de marmol blanco en la misma sala hay el baño de la reina; despues sigue la sala de la servidumbre y despues la del baño del Infante. Despues entramos and la sala donde los reyes reposaban despues del baño. Hay allí camas o canapés de marmol, que eran ciertamente en el tiempo pasado cubiertos de tapetes preciosos. en medio de la sala hay una fuente y arriba varios caeres o galerias donde tocaba la musica. Como casi en todas las salas las paredes estan adornadas de inscripciones arabes y el cielo raso en forme de cupola es de madera y muy bien pintado. Atravesamos la sala de las nimfas y entramos en la antesala de la mesquita y por fin en la misma mesquita convertida en capilla por Fernando y Isabel. Bajamos despues a la sala de los Embajadores y antesala de la de los Embajadores. Todas las paredes adornadas de magnificos arabescos de estuco pintados y dorados con un cielo raso de madera de mucho merito. Se dice que Isabel la Catolica recibió en esta sala a Cristobal Colon despues de su primer viaje a America. en la sala de los embajadores hay 9 nichos o alcobas con rejas. Todo el edificio se halla en decadencia, las paredes y muros estan llenas de grietas; sin embargo empiezan ahora poco a poco a restaurarlo. Es patrimonio de los reyes. Los techos son comunes y consisten en tejas. Fuimos despues a la Cartuja que esta extramuros que es una joya de obras maestro. en ella hay una gran galeria de cuadros pintados por un fraile cartujo, que representan los suplicas a los cuales furon somisos – sometidos los cartujos de Inglaterra que no querian obedecer a Enrique VIII y reconocerlo como cabeza de la iglesia. en otras salas hay pintadas por el mismo columnas de marmol y una cruz que parecen ser naturales y so-

[66]
lamente acercandose se puede reconocer la ilusion. en la iglesia hay una abundancia de marmoles rojos hermosisimos, cuyos venas muchas veces representan hombres; tambien marmoles pardos. Los atriles, vasos de aqua santa, altares, todo es de marmol magnifico; los muros cubiertos de estuco un poco a la arabesca; los cielos rasos de magnificos afrescos. Las puertas son de ebano y cubiertas de conchas, de marfil y nacar de mucho merito. Allí hay una multitud de cajoneras de ebano, riquisimamente adornadas de nacar, de marfil y de concha del siglo XVI. Volvimos a la fonda. Mi guia un hombre sin ninguna educacion pero poseido – endemoniado de mucho orgullo, no queria ir mas lejos conmigo y pidió 5 francos para sus servicios. Yo le hice entegrar 2. Despues del almuerzo fui al generalife o casa de campo de los reyes moros. Habia de seguir mas arriba la bellisima alameda que conduce a la Alhambra. en la casa del generalife hay 2 salas con retratos de reyes españoles y algunas inscripciones arabes; por la abundancia de las aguas con las que se riega los jardines esos son magnificos y en forma terraplenes o terrados los unos sobre los otros. Fui despues a los jardines del banquero madrileño Calderon, que estan muy bien mantenidas. Los arboles no tienen que pocos años pero ya llevan fruta deliciosa. Hay un estanque artificial con un alto puente encorvado. en medio del estanque hay una imitacion de un burgo morisco destruido por el tiempo. Sobre el estanque nadan 2 cisnes. Hay una cascada, algunas estatuas y hibernaculos o invernadores de piñas, que se venden aqui por 3 y 4 $; mitigué mi sed con peras y uva y salí. Apenas hube
[67]
hecho algunos pasos cuanto encontré a un caballero, que me pareció hombre de bien y con quien entablé conversacion. El me dijo que iba a la iglesia de San Miguel extramuros y acep-

té con buena gana su proposicion de acompañarme allí porque es hoy el dia de San miguel y hay allí una multitud inmensa de personas de ambos sexos. Habiamos de subir un monte alto pero estabamos mucho recompensados por la vista de las guapisimas señoritas que se apresuraban a montar – trepar. Arriba habia algunas tiendas de tela lienzo and las que se vendian refrescos. Volví a mi casa y llegue en tiempo para la comida.

Granada sabado el 1 Octubre. Segun o conforme a su promesa vino ayer por la mañana a las 10 almorzar conmigo el librero Carlos gomez Moreno, que me acompaño despues a la Catedral edificio suntuoso del siglo XV adornado con muchos marmoles de gran merito y de los sepulcros de los reyes catolicos Fernando y Isabel y de Felipe I y de su reina Juana. Fuimos despues a la Alhambra, donde recorrí todavia una vez todas las suntuosas salas y patios. Estan ahora restaurando aquel palacio y en verdad ya es tiempo porque ya se ha borrado el color casi en todas partes y las grietas en las paredes amenazan con pronta destruccion. El librero se fúe despues vaguear – vagar a sus ocupaciones, mientras yo me fúi a mi paisano Pablo Nottbeck, que es aqui muy popular tanto por los gastos que hace, que por su afabilidad. El me recibió con mucho agasajo y me enseñó mostro las obras maestro ↓de imitacion en miniatura↓ que ha hecho de la Alhambra, y que tiene la intencion de exhibir en la exhibicion en Nueva York y de presentar despues al Emperador de Rusia, el cual no faltará de cubrirlo de decoraciones que merece por tan sublime trabajo que exige a la vez un arquitecto y un pintor de súmo mérito. El me condújo a un encuadernador en cuya casa tiene tambien un pequeño taller y era tan amable de acompañarme
[68]
despues a la fonda de los siete suelos, cerca de la Alhambra, donde me forzó comer a sus gastos y no permitiendome pagar siquiera mi parte. En el curso de la conversacion le dije que ya un año y medio habia buscado de balde en mis viajes la ocasion de emplear mi capital con ventaja.

El me respondió que sabia una buena ocasion
teniendo el un amigo colmo de engenio y inteligencia
que habia comprado aqui una fabrica de papel so-
bre el rio de las aguas blancas cerca de Duda
que dista 1½ legua de aqui, que ese sabio se
llamaba Fernando Wilhelmi y que ya sacaba una
ganancia mensual de 60/m rs de su fabrica, en
la que habia introducido muchas nuevas invencio-
nes de su discubrimiento; que el (Nottbeck) habia
en vano procurado – intentado persuadir a sus her-
manos ofrecir a aquel hombre una recompensa –
remuneracion suficiente para empeñarlo que se vaya
a San Petersburgo ameliorar la fabrica de papel que
tienen allí que (segun él) ya da 3000 rubles men-
suales de ganancia y facilmente pudiera dar lo doble
por fin que el hombre le habia dicho que este era
el pais para la remolacha y que una fabrica de
azucar se haria pagada aqui en 2 años, porque
la remolacha crecia aqui todo el año. Alegréme
mucho de semejante persfectiva y exprimí –
manifesté el deseo conocer al Wilhelmi. Nottbeck
(D^{on} Pablo) ofreció acompañarme allí a caballo.
Salimos pues y alcanzamos la fabrica en una
hora y media. El esterior del edificio con su alta
chimenea me hizo creer en la importancia
de la fabrica. Wilhelmi, aunque no hablia bien
siquiera su propio idioma, me hizo la impresion
de un honrado fabricante aleman, que no carece
conocimiento en su arte, pero en la conversacion
que tuvo con el no alcanze descubrir nada de
aquella superior inteligencia, que tanto me
habia ponderado Nottbeck. Aun descubrí
luego que él tenia ideas encaresidisimas –
ideas sumamente encarecidas de aquella empresa
y sus repetidas contradicciones, porque luego
me dijo que una fabrica para trabajar 5/m quin-
tales de remolachas costaria fr 400/m y despues que
por este precio se pudiera construir una que
trabaja 200.000 quintales de remolachas y a razon
[69]
de 7% de azucar 14.000 quintales de azucar, igual
cantidad de espiritu de vino, gran cantidad de
jarab – almibar y de estércol – fiemo. Me
dijo que para la leña bastarian los arbustos espi-
nosos que con pocos gastos se procura de los montes

y que los cultivadores se empeñarian con alegria al trabajo de cultivar las remolachas si se les diese solamente la semilla y les garantiria la venta a precios fijos. El añado que segun su parecer el mejor lugar para el establecimiento seria al rio y cerca del pueblo de Monachil que dista 1½ de aqui, porque allí ayudario la corriente del agua que hace allí una cascada de 20 piés.
Nos hospedó miserablemente; habia de acostarme en el olor del comun sobre un sucio colchon y almohadas todavia mas sucias. No podia adormecer por las mordeduras de las muchas chinches y pulgas y cuando me adormecí fúi de vez and cuando despertado por el perro que se acosto sobre mi colchon. Apenas podiamos obtener esta mañana un vaso de agua; no era cuestion de refrescos porque no habia absolutamente nada que ofrecirnos. Ví la fabrica, que esta movido por el agua y es tan pequeña que me parece imposible que pudiera trabajar para mas de 30/m $ de papel todo el año y no puede ser grande la ganancia aunque paga solamente 1500 $ de arrendamiento y compra barato los harapos.
Volvimos hambrientos y sedientes, pagué 48 rs para los 2 caballos, almorzamos juntos y nos separamos. El camino de aqui a la fabrica conduce continuadamente sobre el declive o el pendiente de los montes y abajo se vé el hermoso valle regado por un arroyo llamado rio por falta de otros mayores. Tomé aqui algunas informes cerca del jardinero de Calderon y el librero moreno y supe luego que el establecer una fabrica de azucar de remolachas seria una empresa ruinosa luego que los cultivadores no aconsentirian cultivar a remolachos que chupan o consuman la fertilidad de su termino y dejar de cultivar las hortalizas que estan acostumbrados a cultivar; que aun si ellos proporcionasen la cantidad – copia necesaria de remolachas estas no podrian sostener la
[70]
competencia de la caña azucar, que se produce sobre la costa en tanta abundancia que se vende a 4 rs la arroba y que la falta de la leña prohibiria de una vez la ejercicion da tal atre-

vido proyecto. Mucho me ha gustado Nottbeck que es un subdito escelente y tan honrado y sincero que cree los otros incapaces de mentir y cree todo lo que dicen. Por supuesto él ya ha sido engañado muchos veces.

Malaga Martes 4 Octubre. Salí de granada Sabado el 1 Octubre a las 6 de la noche por la diligencia; Tuve un asiento en el coupé por lo que pagué 70 rs. Estuvimos alli 4 pasajeros. A mi lado estaba asentada una joven muy linda de 17 años natural de valencia, que me parecia al principio persona de bien. Pero luego reconocí por la libertad de sus ademanes y gastos que habia de ser puta. Tuve lastima y compasion con esa pobrecita, que probablemente ha sido echado en este vida desdichosa por la culpa de sus parientes o por un vil seductor. Ella era muy guapa con un nariz aguileña, pequeña boca y pelo negro. Las mejillas – carrilos pintados con color rojo me confirmaron and mi idea de su condicion.

Cuando no podia dormir en su rincon por la oscilacion y los traqueos, las sacudidas del coche yo la tomé en mi brazo y inclinada su cabeza sobre mi hombro ella dormio varias horas excitandome a tal punto que no podio discurrir a dormir. Debilitado y aflojado por mi dulce carga llegué por fin el Domingo 2 corr. a las veras de Malaga donde ya la nueva mama aguardaba la chiqua con un coche para conducirla S. Antonio N° 13 o 17. Tuve la intencion de pasar la noche con ella, pero oyendo aqui en todas partes que ya estaba esperada su venida de muchos y que al menos una docena se habia arrojado sobre ella el mismo dia, sin permitirla a descansar, y habiendome dicho un compañero de viaje que ya la habia dado dos
[71]
veces pruebas de su amor, tuve miedo y me abstuve de ella. Durante el tiempo que yo faltaba de aqui – que hacia falta de aqui la temperatura no se ha destemplado y parece que ha todavia aumentado el calor. Por desgracia no hay vapor hasta mañana por

la noche cunado saldré con el "Egyptien"
con direccion para Alicante. Pasaje 13$.
El Domingo por la noche en el paseo de la
alameda quise asentarme sobre una silla cuando
un hombre que estaba asentado cerca [dijo] que ya
estaban alquiladas las sillas, pero me convido
de tomar una silla hasta que llegasen sus niñas.
Habiendo oido de mi que era ruso él manifesto
gran deseo conocerme mas y me presentó sus
hijas de quienes la una tiene 13 y la otra poco
mas o menos 20 años. Hablé con ellas benig-
namente y a su convite les prometí que
vendria a verlas ayer, lo que executé con
buena gana porque estaba curioso a ver como
viven aqui las familias. Encontrélos a la mesa
propuse tomar un palco and el teatro y volver
cuando habrian comido. "Que disparate" me respon-
dio la hija mayor Cristela. Consintieron con
buena gana. Volví con el billete y mientras
las jovenes se vistieron yo me fui con el padre
a un cafe. Fuimos despues juntos al teatro.
Dí el brazo a la hija mayor. El palco
era de segunda rango ↓30m r y 16 de entrada↓ y no me gustaba. Tuve
alguna verguenza de haberlo tomado y sentí
de no haber pagado un duro mas y tomado
primer rango. Representaran: "mis 2 mujeres"
con cantos y buena musica y segun mi parecer
se quedaron contentas las niñas. Me acom-
pañaran a mi casa despues del teatro.
El teatro es pequeño y apenas cabe siquiera
400 personas. La entrada cuesta de 4 hasta
5 ^{r}; los sillones en la platea poco mas o
menos de ½ duro. Nunca me hubiera ima-
ginado que españoles, que hacen pretension
[72]
a ser gente decente, puedan vivir tan mi-
serablemente como esta familia vive. Parece
que no tienen mas de dos cuartos muy pequeños
uno abajo otro arriba y en el mismo cuarto
abajo tiene su taller un cerrajero italiano.
Las niñas cuidan de la casa, cumplen con to-
dos los trabajos sucios de la criada y por eso
no pueden estar muy limpias como tam-
bien lo demuestran sus manos y uñas. Su in-
teligencia esta apenas a la altura de sus

ocupaciones, y, como parece evidente por las respuestas que dieron, ellas no tienen siquiera la minima educacion. Pero la falta de esa ellas substituyen por su gracia y por sus ojos amorosos. A mi pregunta si la minor hija era todavia en la escuela respondio el padre que un maestro venia instruirla en casa, que no podia mandarla a la escuela porque ella estaba siempre proseguido de muchachos de su edad que apresuraban hacerla el amor y goderla. El añadió que las pasiones estan aqui mucho mas ardientes y fuertes en ambos sexos. Hoy por la noche hay musica en la alameda. No encontré alli a mis amigos y iré a su casa mañana.

Jueves el 6 Octubre a la altura de Almeria a bordo del vapor frances "Egyptien" con direccion de malaga para Alicante. Me despedí ayer del Sr Matas y de sus niñas que me recibieron con gran regocijo porque ya pensaban que me habia marchado sin despedirme de ellas. Mostraronme sus camerotes pequeños y me enseñaron unas hermosas flores artificiales mezcladas con glandulas. Compré de ellas por 5$ y entregué las flores a las niñas diciendo que las lleven en recordacion de mi. Ofrecieron mandarme a Paris el retrato de la hija mayor.
Me apresuré disuadirlas de semejante disparate ↓puro↓. Les dije que su retrato resteria siempre grabado y imprimido en mi animo – alma y no quieria por los peligros del viaje. Las dejé sumamente contentas y en la esperanza de que volveria por aqui en el invierno. Hube de complir con varias formalidades
[73]
acerca de mi pasaporte, porque el armador del vapor frances no queria darme el billete antes que entregase el pasaporte refrendado por la Sanidad y esa no queria refrendar antes que fuese refrendado por la policia. Como suponia que esta se apondria a refrendar hasta que hubiese logrado la refrendacion del consul ruso me fue luego a este, que se llama Rey y cuyo agente a Petersburgo esta Ponfick; despues la hice refrendar por el consul frances, despues por la policia y por fin por la Sanidad, tomé mi billete por lo que pagué 13$ y me fúi a bordo despues de la comida. Las disposiciones a bordo son muy malas, no hay malos camarotes y

peores camas sino para 12 pasajeros y como hay 15 hemos de dormir nosotras 3 sobre los bancos. Todo es sucio aqui y no se vee que estamos a bordo de un vapor frances.
Alicante Viernes el 7 Octubre. Estoy muy satisfecho de mi viaje de Malaga por aqui. en verdad he raramente disfrutado un viaje tanto como este. Tenia muy buena sociedad de jovenes naturales de Tenerifa que iban al colegio de madrid y entre ellos el marques de la Florida, joven de 22 años hombre de profundos conocimientos de geografia y historia, cuya conversacion me gustaba mucho.

Me recomendó mucho y me encargó que si acaso yo me dirije a la isla de Tenerifa de hacer una visita a su madre en Santa Cruz de Tenerifa o en su villa de la Orotaña, lo que haré sin falta. La buena compañia que túve me indemnizó bastante por las incomodidades del vapor, que estaba arreglado mas para flote que para pasajeros; no habia camarote para mi y debia acostarme sobre las almohadas de los bancos. Llegamos al puerto de Alicante esta mañana a las 6, pero habiamos de esperar hasta las 9 para desembarcar, porque el gobierno en su solicitud para el bien publico habia declarado Alicante: "sucio". Pero no puedo meter esa medida en acuerdo con lo que debiamos alzar la bandera de la cuarantena y que nos hicieron dificutad and el desembarcar. Por fin desembarcamos. Registraron el equipaje a la aduana y fuimos a la fonda del Vapor. Vestime y me daron abundante almuerzo mejor que he encontrado and España hasta ahora. Me fui despues a la casa del Sr Pa[t]ricio de Satrustegui para informarme cerca de su madre sobre la suerte de su hijo minor Miguel. Pa[t]ricio de S. es compañero y socio de Sr Lopez a la Habana y director de los vapores ↓de correo↓ que este ha establecido aqui en España.
[74]
El es acreditado aqui como hombre de mucho entendimiento y se dice que conduce sus negocios bien. No hallé en casa sino la hija minor Bernarda joven soltera de poco mas o menos de 30 años, que me recibió con mucha afabilidad. Ella me gustaba mucho, porque manifestaba tanta modestia y amabilidad; tanta franqueza y altivez de caracter. En contradiccion con lo que me habia dicho a San

Sebastian su hermano Joaquin ella me dijo que Miguel se habia casado a la Habana con una joven americana pobre y se habia establecido a Portland con un español que habia dado el dinero. Despues de su salida (partida) de Sonora, sobre cuyo motivo guardaba un silenzio misterioso, exprimiendo por sus facciones de cara algun embarazo, habia atravesado el Mejico y se habia hecho vice consul ingles en Cardenas cerca de la Habana; habia obtenido este empleo, por haber nacido and Inglaterra y por ser pues subdito ingles, en edad de 22 años, pero habia tenido la dicha diguidad solamente 2 años. Su hermana Ventura estaba con la madre en Madrid, de donde volverán solamente mañana por la mañana a las 10. Tambien encontré aqui la hermana Inez casada a Perfecto Emanuel de Olalde secretaria del gobierno civil y residente ahora a malaga. Esta Señora me parece todavia mucho mas sencilla que Bernarda. Tienen buen alojamiento, al menos para Alicante.

Barcelona Lunes 10 Octubre 1859

Me despedí de Alicante el Viernes a las 5 ¾ horas de la tarde por el ferro carril hasta Almansa, de allí por la diligencia poco mas o menos de 7 leguas y despues de nuevo por el ferro carril. Atravesamos un hermoso pais particularmente en los contornos de Valencia. Allí hay en todas partes abundancia de agua y la vega es riquisima en hortalezas, bosques de palmeras, naranjos, ↓quindos, ciruelos↓ y otros arboles fructiferos. Llegamos a Valencia el Sabado a las 9 de la mañana. Paréme en la fonda del Cid. Salí para ver la ciudad, la cual es tan diferente de las otras ciudades de España como el caracter y el tipo de los valencianos es diferente de otros pueblos. Las calles son estrechas (angostas) y muchas no estan empedrados.

[75]

las casas por la mayor parte grandes y bien construidas. El tipo del pueblo se asemeja mas al dinemarques que al español y no hé acertado descubrir ninguna mujer hermosa. Pero a la primera vista se vee que las mujeres piensan mas al trabajo y arreglo de su casa que al adorno de su cuerpo.Negligencia y descuido

como en Andalucia no se vee en Valencia and
niguna parte. Se dice que hay allí una casa
de putas, en la que las doncellas estan registra-
das diariamente 2 veces por facultativos; no ob-
stante el gobierno no se carga de ninguna ga-
rantia. Salí a las 2 de la tarde por el vapor
Mercurio con direccion para Barcelona donde lle-
gamos ayer a la 1 de la tarde. Debi-
amos esperar 2 horas hasta que nos permitian
de desembarcar. Registraron de nuevo mi equipaje.
Tomé un muchacho que me trajó el equipaje a la
fonda del Oriente, donde obtuve una camara sobre la
calle, que es la hermosa alameda llamada Rambla.
Era Domigo y todas las calles estaban llenas de
gente endominguerada. Mucho me gusta la ciudad
que compite – rivaliza con Paris en hermosura
y solidez de construccion. Aqui todo es calculado
por ganancia y ventaja y las casas estan con-
struidas tan altas para la mayor renta. Arriba
las casas son platas para secar la ropa blanca
sobre los terrados. Hay todavia un otro paseo el
cual tiene 700 varas de larga por 60 de ancho. Em-
pieza en el jardin del general, que es muy hermoso
y lleno de naranjos y pimenteros, que tienen de
una vez flores y fruta. Las veredas son magnificas.
El tipo de la poblacion me parece puro gotico; al
menos no reconosco ningun vestigio de raza arabe o
romana aqui. Con raras excepciones las mujeres son
muy feas aqui. Estúve ayer al teatro principal
que es mucho mas grande que el nuestro de michel [...]
y hoy por la noche estuve en el Liceo, que
se dice ser un poco mas grande todavia que la Scala
de milan. Representan "Il trovatore". Por desgra-
cia la pieza; el liceo es superbio; todos los sillones
estan cubiertos de terciopelo rojo. Solamente no hay
lustre; las decoraciones son groseras y los actores
y actrices – cantores y cantoras estan lejos de ser de
[76]
primer talento. Durante los entreactos tuve
una conversacion graciosisima con una Señorita
de nombre Chasq que estaba asentada detras de mi.
Ella parecia bastante instruida para una española
y mas linda que he visto aqui hasta ahora.
Me disgusta la aristocracia de la nobleza y hidal-
guia pero mucho me gusta la aristocracia de

la hermosa española. Tomé hoy por la maña-
na un cicerone que me condujo en todas partes
no se debe buscar aqui antiguedades. El
ingenio trabajador y emprendador de los catalanes
cuida poco los monumentos de la antiguedad
y se derriban los antigüos edificios para que
den lugar a construcciones modernas de utili-
dad. Lo unico que pueda interesar aqui un ar-
cheologo es el archivo aragones en el que me
hizo entrar mi curiosidad. Hay alli mas
de 600 bulas pontificias de pergamento en rollos
con grandes sellos, ademas cuantidad inmensa
de cartes reales y otras escrituras desde al año
1000, todo bien compuesto en libros y encuadernado
en cuero de puerco – cocino. Al salir hice el co-
nocimiento del inspector de los archivos Antonio
de Rofarull, que me acompaño al museo, en el
que no hay nada ademas de unos pocos estatuas
mutiladas. Yendo con el por las calles encon-
tramos un amigo suyo de nombre Juan Guitara
que publica aqui un periodico, y me trajo 2 hojas
inmensas de su album para que me eternizasa
en ellas por un discurso en 16 lenguas sobre
la pipa y el fumar. Como la demanda era esor-
bitante no escribí nada. Las calles de Barcelona
estan muy bien empedradas de piedras platas.
<u>A bordo del vapor Barcelona a la altura</u>
<u>de San Sebastian Martes el 11 Octubre.</u>
Salí hoy a la una de la tarde por el vapor
a tornillo o a helice Barcelona. Muy penoso
era para mi separarme de España donde he
pasado casi 2 meses muy agradablemente.
Tenemos mas como 15 pasajeros y entre ellos
un joven español de nombre Emanuel Augustin
[77]
Calvo natural de la Habana, que va a Amberes
para entrar allí al colegio; se quedara allí un
año y entrará despues en los negocios de su tio
a Sevilla, que no esta casado y no tiene herederos
mas proximos. Tenemos viento favorable ade-
lantamos bien aunque tenemos mucho flete.

بتربوريخ في بلد المسكوب الاحد الخميس عشر تشرين الثاني
قد خمسه جماعت انا يكون هنا __ انا رجعت من الاخر كبير سفري
انا ظنت تصار ___ تخري الشتا في بريش منشان تفهم قرات في

مكتب الكبير مسمي ونيورسيتت __ اما انا لقيت عند روتشيلد مكابيت
الذين دعولي بالمجل الي بترسبوريخ فان في الغيبه ي واحد سريق صار
بدي صلني عر نزعات وانا نخسري نزعات ان ما عجلتش هنا انا __
حالا انا كسبت يحرق من الشوق تشاف __ تنظر الكبير رجل الذي من انا
حتي في القدس كيف حال الاهل ك الزوج ك والاولادك انا كتبتك من
مدينه الروم بعدين انا زارت بر المصر اسكندرية ومصر القاهرة استكريت
مركب مع 8 مركبية واحد مدقه 1 ريس 1 ترجمان 1 خدام و
انا طلعت بحر النيل حتى الثني شلال بعدين انا رجعت الي المصر
شفت في السفر كل البرابي اشتريت حصن جمعتني __ لزقتني مع
جرافين ومع 13 جمال عبرنا البريه __ صحاره حتي القدس

[82]

Madrid		13		
Cadiz		21	James Ferrier Montreal	
Gibraltar		24		
Gibraltar		24 g.	[Mr.] Sterky y hijo	St Petersburgo
ditto	“	Kuchseynsky ditto		
ditto	“	Const Fehleisen ditto		

List of figures

1 Gennadius Library Athens (digital copy from: http://www.ascsa.edu.gr/index.php/archives/schliemann-photos-series-j)

2 Bölke, Wilfried 2016 Dein Name ist unsterblich für alle Zeiten. Düsseldorf.

3 Bron: Bölke, Wilfried 2016 Dein Name ist unsterblich für alle Zeiten. Düsseldorf.

4 Château de Compiègne – Musée du Second Empire (digital copy from: https://en.wikipedia.org/wiki/Battle_of_Solferino#/media/File:Yvon_Bataille_de_Solferino_Compiegne.jpg)

5 Digital copy from: https://upload.wikimedia.org/wikipedia/commons/4/41/Reina_Isabel_II_de_Espa%C3%B1a.jpg

6 Art Institute Chicago (digital copy from: https://i.pinimg.com/originals/fe/69/0f/fe690f961de8997f579d441c0d07c752.jpg)

7 Collection Höhere Graphische Bundes- Lehr- und Versuchsanstalt, Vienna. (zie ook: Nazarieff 1993 Early Erotic Photography. Köln.) Copy from: https://www.bankier.pl/static/att/69000/2043073_foto1.jpg

8 Library and Archives Canada / MIKAN 3506659

9 Das Buch der Geologie oder die Wunder der Erdrinde und der Urwelt. 1855 Leipzig

10 Wout Arentzen

11 Copy from:http://hanganadolosmalos.blogspot.nl/2011/04/santo-domingo-el-real-madrid-historico.html

12 Philadelphia Museum of Art. Copy from: http://www.philamuseum.org/collections/permanent/103634.html

13 British newspaper archive, copy from: http://www.antiquemapsandprints.com/spain-bull-fight-madrid-accident-to-montes-matador-antique-print-1850-104047-p.asp

14 Copy from: https://i.pinimg.com/564x/8f/24/cb/8f24cb14b2e09539597c9423f70ce987.jpg

15 Copy from: https://upload.wikimedia.org/wikipedia/commons/1/1d/Alcazar_segovia_J_Laurent.jpg

16 Copy from: http://www.arquitecturaviva.com/media/Images/visores/julio_2015/BNEelescorial_0.jpg

17 Atkinson Art Gallery Collection – Southport, copy from: https://artuk.org/discover/artists/burgess-john-bagnold-18291897

18 Copy from: https://blogdruta.com/cartoteca/#jp-carousel-10481

19 Copy from: https://americanart.si.edu/artwork/clearing-storm-gibraltar-81938

20 The Alhambra. James Cavanah Murphy (1760-1814) 1815 The Arabian Antiquities of Spain. Copy from: http://islamic-arts.org/2012/the-alhamra-at-granada/

21 Pateo del Agua. James Cavanah Murphy (1760-1814) 1815 The Arabian Antiquities of Spain.

22 Barcelona. Archivo Fotográfico de Barcelona. Copy from: Bron: http://www.anycerda.org/web/es/arxiu-cerda/fitxa/muralla-de-mar/269

23 Gennadius Library Athens Greece. Copy from: http://www.ascsa.edu.gr/index.php/archives/schliemann-diary-a-4

24 Gennadius Library Athens Greece. Copy from: http://www.ascsa.edu.gr/index.php/archives/schliemann-diary-a-4

Bibliography

Agirreazkuenaga, Joseba, 2012: Trayectoria biográfica de Joaquín Marcos Satrustegui Bris (Donostia-San Sebastián 1817-1885), mediador and el convenio de Bergara, diplomático y contrario a la abolición foral de 1876. *Memoria y civilizacíon* 15, 71-89.

Andrusová-Vlčeková, Galina, 1999: Heinrich Schliemann und seine Russian Familie. *Mitteilungen aus dem Heinrich-Schliemann-Museum Ankershagen*, Heft 6.

Arentzen, Wout, 2011: *Schliemanns dramatische Reise im Jahre 1841*. Utrecht.

Bloedow, Edmund F., 2001: Heinrich Schliemann and James Ferrier: A Noteworthy Side-Trip to Canada. *Cahiers des Études Anciennes* XXXVII. Québec.

Bogdanov, Igor, 1998: *Ne privozi s soboju Gomera…* St. Petersburg.

Bogdanov, Igor, 2002: *Peterburgskaja familija: Latkiny*. St. Petersburg.

Bölke, Wilfried, 1996: *Heinrich Schliemann. Ein berühmter Mecklenburger*. Schwerin.

Bölke, Wilfried, 2015: *„Dein Name ist unsterblich für alle Zeiten". Das Leben Heinrich Schliemanns im Briefwechsel mit seiner mecklenburgischen Familie*. Duisburg.

Einsle, Hans und Bölke, Wilfried, 1996: *Das Heinrich Schliemann-Lexikon*. Bremen-Rostock.

Eliot, George, 1859: *Adam Bede* New York.

Ehrenreich, Barbara & **English**, Deirdre, 1979: *For her own good. 150 Years of the Experts' Advice to Women*. London.

Finn, James, 1877: *Byways in Palestine*. London.

Hempel, Rudolph, 1866: *Bericht über die Festlichkeiten bei der Gedenkfeier des 25-jährigen Bestehens von Dr. Krauses Lehr- und Erziehungsanstalt*. Dresden.

Jähne, Armin, 1999: Schliemanns Abschied von Rußland. *Mitteilungen aus dem Heinrich-Schliemann-Museum Ankershagen*, Heft 6, 1999.

Jähne, Armin, 2016a: Archäologie zu Schliemanns Petersburger Zeit. *Mitteilungen aus dem Heinrich-Schliemann-Museum Ankershagen*, Heft 10/11.

Jähne, Armin, 2016b: Archäologie zu Schliemanns Petersburger Zeit. *Mitteilungen aus dem Heinrich-Schliemann-Museum Ankershagen*, Heft 10/11.

Lang Neil, C,. compiled and edited by [ca. 1902]: *Pictorial Adventure and Travel In all Parts of the World*. London.

Ludwig, Emil, 1932: *Schliemann Geschichte eines Goldsuchers*. Berlin-Wien-Leipzig.

Mai, Joachim, 1997: Zur rechtlichen Situation Heinrich Schliemanns in Russland 1846 bis 1864. Heinrich Schliemann und seine russische Familie. *Mitteilungen aus dem Heinrich-Schliemann-Museum Ankershagen*, Heft 5.

Mai, Joachim, 1999: Heinrich Schliemanns sozialer Aufstieg in Rußland 1846 bis 1864. *Mitteilungen aus dem Heinrich-Schliemann-Museum Ankershagen*, Heft 6.

Meyer, Ernst, 1953: *Heinrich Schliemann Briefwechsel*. Berlin.

Meyer, Ernst, 1969: *Heinrich Schliemann Kaufmann und Forscher*. Göttingen.

Schliemann, Heinrich, 1881: *Ilios the City and Country of the Trojans*. New York.

Schliemann, Sophie, 1892: *Heinrich Schliemann's Selbstbiographie*. Leipzig.

Stoll, Heinrich A. (Ed.), 1986: *Abenteuer meines Lebens. Heinrich Schliemann erzählt*. Leipzig.

Stucky, Rolf A., 2016: „Wollen Sie Schliemann sehen?“. *Mitteilungen aus dem Heinrich-Schliemann-Museum Ankershagen*, H. 10/11.

Thanos, Christo & **Arentzen**, Wout, 2012: *Without having seen the Queen*. Leiden.

Thanos, Christo & **Arentzen**, Wout, 2014: *Schliemann and the California Gold Rush*. Leiden.

Tomalin, Claire, 1990: *The Invisible Woman: The Story of Nelly Ternan and Charles Dickins*. London.

Traill, David A., 1993: *Excavating Schliemann*. Illinois Classical Studies, Supplement 4. Atlanta.

Traill, David A., 1995: *Schliemann of Troy. Treasure and Deceit*. London.

Tulschinsky, Gerald J.J., 1982: Ferrier, James. *Dictionary of Canadian Biography* vol 11. Toronto.

Index

T

U

V

W

Z